Fire Service Pump Operator's Handbook

Warren E. Isman
Fire Chief, Fairfax County (VA) Fire and Rescue Service

A PennWell Publication

Dedication

The writing of a textbook takes a great deal of time during which many chores get neglected. This neglect is hardest on a family. So, for their understanding and help during this hectic period I am dedicating this book to my wife Marguerite, and my children Emily, Susan, Amy, Kenneth, Matthew, Janet, David, Carla and Theodore.

The Standard for Fire Apparatus Driver-Operator Professional Qualifications, Chapter 3, NFPA 1002-1982, copyright 1982, National Fire Protection Association, Quincy, MA 02269, reprinted with permission. This reprinted material is not the complete and official position of the NFPA on the referenced subject, which is represented only by the standard in its entirety.

Copyright 1984, 250 Fifth Ave., New York, N.Y. 10001. All rights reserved.
No part of this book may be reprinted by any process whatsoever without permission in writing from the publisher.

Printed in the United States of America

Library of Congress Catalog Card Number 84-080977
ISBN 0-87814-905-8

Other Related Fire Service Books Available From Fire Engineering.

The Fire Chief's Handbook, *Casey*
Fire Fighting Principles & Practices, *Clark*
High Rise Fire & Life Safety, *O'Hagan*
Fire Service Hydraulics, *Casey*
Fire Service Hydraulics, A Study Guide, Questions & Answers, *Sylvia*
Flammable Hazardous Materials, *Meidl*
Hazardous Materials, *Isman & Carlson*
Fire Fighting Apparatus & Procedure, *Erven*
Introduction to Fire Prevention, *Robertson*
Fire Suppression & Detection Systems, *Bryan*
Fire Dept. Management: Scope & Method, *Gratz*
Arson Investigation, *Carter*
Introduction to Fire Science, *Bush & McLaughlin*
Fire Dept. Operations with Modern Elevators, *McRae*
Strategic Concepts in Fire Fighting, *McAniff*
Investigating the Fireground, *Phillipps & McFadden*
Modern Suburban Fire Fighting, *Sylvia*
Fire Service Pump Operator's Handbook, *Isman*
Managing Fire Services, *Bryan & Picard*
Fire Fighting Operations in Garden Apartments, *Gaines*
Hazardous Material Emergencies, *Cashman*
Emergency Rescue, *Erven*
Practical Approaches to Firefighting
Vehicle Rescue, *Grant*
Winning the Fire Service Leadership Game, *Caulfield*
The First Responder in Action, *Nixon & Brown-Nixon*
The Common Sense Approach to Hazardous Materials, *Fire*
The Common Sense Approach to Hazardous Materials Study Guide, *Fire*
Introduction to Fire Apparatus and Equipment, *Mahoney*
Introduction to Fire Apparatus and Equipment Study Guide, *Mahoney*
First Responder, *Bergeron*
Emergency Care, *Grant, Murray & Bergeron*
Fire Service Directory of Training and Information Sources
Your First Response in the Streets, *Briese & Schottke*
Collapse of Burning Buildings: A Guide to Fireground Safety, *Dunn*
Fire Service Radio Communications, *Spahn*
Fire Instructor's Training Guide, *Bachtler*

Preface

Today's pump grew out of a basic need—the need to get water on fire as quickly and efficiently as possible.

The advancements made through the centuries dictated the need for a pump operator, which has become a full-time specialization. With today's highly sophisticated equipment, a fire apparatus driver/operator must meet certain performance standards set forth by the National Fire Protection Association.

This book provides the information necessary to efficiently operate a fire department pumper and to meet the requirements set forth in Chapter 3 of the NFPA 1002 Standard, Fire Apparatus Driver/Operator Professional Qualifications 1982, "Apparatus Equipped With a Fire Pump."

Starting with the basics (addition, subtraction, multiplication, division) and working through fractions, ratios and hydraulics, this book discusses the different types of pumps; the equipment carried on or attached to a pumper and their testing and maintenance; nozzle reactions; pressure control systems; priming devices; the different operations for drawing water from various sources; and driver training.

In the interest of providing a step-by-step progression, not all the paragraphs in NFPA Standard 1002 are dealt with in the order set forth in the standard.

Water movement, flows, and a review of equations and definitions are covered in the appendices at the back of the book.

Contents

Chapter		Page
	Preface	
1	Basic Mathematical Operations	1
2	Fractions	9
3	Algebra	24
4	Ratio, Proportion, Percentage, Powers and Roots	33
5	The Metric System and Fire Service Hydraulics	43
6	Fireground Hydraulics	47
7	Introduction to Pump Operations	75
8	Positive Displacement Pumps	80
9	Centrifugal Pumps	90
10	Pump Drives	117
11	Pump, Cab, Body Components	128
12	Nozzle Reaction	141
13	Pressure Control Systems	144
14	Priming Devices	169
15	Drafting Operations	182
16	Hydrant Supply and Operations	189
17	Relay Operations	201
18	Sprinkler and Standpipe Operations	209
19	Tanker and Portable Pump Operations	225
20	Testing and Maintenance	234
21	Driver Training	247
Appendix		
A	Understanding Fluids	252
B	Water Movement	256
C	Quantity of Water Flowing	265
D	Equations	271
E	Definitions	272
	Index	277

Chapter 1

Basic Mathematical Operations

Many firefighters, faced with the opportunity of becoming a pumper/driver, become concerned because of the calculations that will be necessary to operate the pump correctly. Firefighters who have been out of school for some time are afraid that they will not remember some of the basics of mathematics. For those who are worried, this chapter is presented as a review. Read it carefully, for it provides the foundation for all of the calculations in this book.

There are two common systems that can be used to express numbers—Arabic and Roman. The one most commonly used, and the one more familiar to firefighters is the Arabic system. The Arabic system is based on 10 figures called digits: 0, 1, 2, 3, 4, 5, 6, 7, 8, and 9. By using these digits in various combinations, any number can be expressed.

The value of the figure is dependent upon the location of the digit within the number. A mark, called a decimal point, is used to separate whole numbers from parts of a whole number. Whole numbers are written to the left of the decimal point, while parts of a whole number are written to the right of the decimal point.

The place names for the digits are:

MILLIONS	HUNDRED THOUSANDS	TEN THOUSANDS	THOUSANDS	HUNDREDS	TENS	UNITS	DECIMAL POINT	TENTHS	HUNDREDTHS	THOUSANDTHS

Using this system, the number 1296317.845 would be read as one million, two hundred ninety-six thousand, three hundred seventeen and eight hundred forty-five thousandths.

1

Another important part of mathematics is the shorthand symbols that are used to indicate the mathematical operations to be performed. The symbols used for this text are:

Symbol	Word	Meaning	Example
+	plus	addition	$5 + 7 = 12$
−	minus	subtraction	$7 - 5 = 2$
×	times	multiplication	$5 \times 7 = 35$
÷ , /	divide	division	$6 \div 3 = 2$
			$10/5 = 2$
=	equals	equality	$3 + 2 = 4 + 1$
( )	parentheses	perform the operation contained within	$(3 + 2) \times (4 + 1) = 5 \times 5 \times 25$
( ) ( )	parentheses	multiplication	$(3 + 2)(5 - 3) = 5 \times 2 = 10$

ADDITION

Addition is combining two or more numbers to yield one answer, the sum. The steps for addition are:

Step 1. Write the numbers one under each other, with all the unit digits in the unit column, the tens digits in the tens column, the hundreds in the hundreds column, etc.

Step 2. Add all the digits in the units column. If the total is greater than 9, write the unit digit of the total in the unit column and carry the other digits to the tens column.

Step 3. Add all the digits in the tens column, including the digits carried from the unit column. If the total is greater than 9, write the unit digit in the tens column and carry the other digits to the hundreds column.

Step 4. Continue to add each column, carrying forward the digits greater than 9.

Example: Four pumpers on the fireground are supplying 526 gpm, 738 gpm, 1226 gpm, and 946 gpm. What is the total amount of water being supplied?

Step 1. Arrange vertically

```
  526
  738
 1226
  946
 ----
  212
```

Step 2. Add the units column. The number 26 is greater than 9 so the 6 is written in the unit column and the 2 is carried to the tens column. 6

Step 3. Add the tens column. The number 13 is greater than 9 so the 3 is written in the tens column and the 1 is carried to the hundreds column. 3

Step 4. Add the hundreds column. The number 24 is greater than 9 so the 4 is written in the hundreds column and the 2 is carried to the thousands column. 4

Step 5. Add the thousands column. 3

                                   3436

BASIC MATHEMATICAL OPERATIONS

The total flow at the fireground is three thousand four hundred and thirty-six gallons per minute.

Addition should always be checked by adding the digits in reverse order. If the original was added from top to bottom, check your work by adding from bottom to top.

Two rules for performing mathematical operations of addition are:

Rule 1: If a and b are used to represent any two numbers, then it can be said that $a + b = b + a$. This is called the *commutative* rule of addition and means that the order of addition is reversible.

Rule 2: If a, b, and c are used to represent any three numbers, then it can be said that $a + (b + c) = (a + b) + c$. This is called the *associative* rule of addition and means that addition can be performed in any sequence. This is the rule that permits addition to be checked by adding the digits in the reverse order.

MULTIPLICATION

Multiplication is a simplified process of adding the same number a given number of times. For example, 15 added together three times will equal 45. This can be accomplished either by

```
   Addition          or       Multiplication

      15                            15
      15                         ×   3
      15                           ―――
     ―――                            45
      45
```

The terms used for multiplication are:
$$\text{Multiplicand} \times \text{Multiplier} = \text{Product}$$
In performing multiplication operations, it does not matter which number is written first, the product is always the same. Thus, you may write $15 \times 3 = 45$, or $3 \times 15 = 45$. (Refer to rule 3 later in chapter.) For ease of operations, however, it is usually best to use the smaller number for the multiplier. The steps for multiplication are:

Step 1. Place the multiplier below the multiplicand, with the unit digits in the unit column, tens digits in the tens column, etc.

Step 2. Use the unit digit of the muliplier and multiply the unit digit of the multiplicand with it. If the resultant number is greater than 9, write its unit digit in the units column and carry over the other digits to the tens column of the multiplicand.

Step 3. Use the unit digit of the multiplier and multiply the tens digit of the multiplicand with it. Add the carryover from step 2. If the resultant number is greater than nine, write its unit digit in the tens column and carry over the other digit to the hundreds column of the multiplicand.

Step 4. Continue this process until all the digits of the multiplicand have been multiplied by the unit digit of the multiplier.

Step 5. Use the tens digit of the multiplier and multiply the unit digit of the multiplicand with it. If the resultant number is greater than 9, write its tens digit in the tens column and carry over the other digits to the hundreds column of the multiplicand.

Step 6. Continue this process until all the digits of the multiplicand have been multiplied by the tens digit of the multiplier.
Step 7. Continue until all digits of the multiplier have been used.
Step 8. Add the resultant figures in each column.

Example: If a particular pump can deliver 248 gpm, how much water can the pumper supply in 23 minutes?

Step 1. Arrange multiplier and multiplicand:

$$\begin{array}{r} 248 \\ \times\ 23 \\ \hline \end{array}$$

Step 2. Use the units digit of the multiplier and multiply the units digit of the multiplicand (3 × 8 = 24). Write the 4 in the units column and carry the 2 to the tens column of the multiplicand:

$$\begin{array}{r} 2 \\ 248 \\ \times\ 23 \\ \hline 4 \end{array}$$

Step 3. Use the unit digit of the multiplier and multiply the tens digit of the multiplicand (4 × 3 = 12). Then add the 2 that was carried over from step 1 (12 + 2 = 14). Write the 4 in the tens column and carry the 1 to the hundreds column.

$$\begin{array}{r} 1 \\ 248 \\ \times\ 23 \\ \hline 44 \end{array}$$

Step 4. Use the units digit of the multiplier and multiply the hundreds digit of the multiplicand with it (3 × 2 = 6). Now add the 1 carried from step 3 (6 + 1 = 7). Write the 7 in the hundreds column.

$$\begin{array}{r} 12 \\ 248 \\ \times\ 23 \\ \hline 744 \end{array}$$

Step 5. Use the tens digit of the multiplier and multiply the units digit of the multiplicand with it (2 × 8 = 16). Write the 6 in the tens column and carry the 1 to the tens column of the multiplicand:

$$\begin{array}{r} 1 \\ 248 \\ \times\ 23 \\ \hline 744 \\ 6 \end{array}$$

Step 6. Continue the process until all the digits of the multiplicand have been multiplied by all the digits of the multiplier:

BASIC MATHEMATICAL OPERATIONS

$$
\begin{array}{r}
1 \\
248 \\
\times23 \\
\hline
744 \\
496 \\
\hline
\end{array}
$$

Step 7. Add the resultant figures:

$$
\begin{array}{r}
248 \\
\times23 \\
\hline
744 \\
496 \\
\hline
5704 \text{ gallons}
\end{array}
$$

Multiplication can be checked by reversing the multiplier and the multiplicand.

In order to be proficient in the multiplication process, the student must memorize the multiplication tables from 1 through 12.

Three rules for performing mathematical operations of multiplication are:

Rule 3: If a and b are used to represent any two numbers, then it can be said that $a \times b = b \times a$. This is called the *commutative* rule of multiplication and means that the order of multiplication is reversible.

Rule 4: If a, b, and c are used to represent any three numbers, then it can be said that $(a \times b) \times c = a \times (b \times c)$. This is called the *associative* rule of multiplication and means that the multiplication can be performed in any sequence.

Rule 5: If a, b, and c are used to represent any three numbers, then it can be said that $a \times (b + c) = (a \times b) + (a \times c)$. This is called the *distributive* rule of multiplication.

SUBTRACTION

Subtraction is the opposite process of addition and is used to determine the difference between two quantities. In performing subtraction, the larger number (the minuend) is placed on top, while the number to be subtracted (the subtrahend) is placed on the bottom. The steps for subtraction are:

Step 1. Write the minuend on top and the subtrahend on the bottom, being sure that the units column, tens column, etc., line up.

Step 2. Start with the units column and subtract the subtrahend from the minuend.

Step 3. Continue sutracting in the other columns.

Example: A pumper is delivering 362 gpm through two lines to the fireground. One line, supplying 251 gpm is shut down. How much water is still being delivered to the fireground?

Step 1. Write the minuend and subtrahend:

$$
\begin{array}{r}
362 \\
-251 \\
\hline
111 \text{ gpm}
\end{array}
$$

Step 2. Start with the units column and subtract the subtrahend from the minuend.

Step 3. Continue the subtraction of the other columns.

Sometimes, however, even though the subtrahend is smaller than the minuend, individual digits of the subtrahend are larger. When this happens, 10 units must be borrowed from the preceding column. This is possible because a number like 481 can be written:

$$400 + 80 + 1$$
$$400 + 70 + 11$$
$$390 + 90 + 1$$

In the second case, 10 was borrowed from the tens column and added to the units column. In the third case, 10 was borrowed from the hundreds column and added to the tens column.

Example: If a pump is supplying a discharge pressure of 192 psi, there is a friction loss of 87 psi. What is the remaining pressure at the nozzle?

Step 1. Write the minuend and subtrahend:

$$\begin{array}{r} 192 \\ -87 \end{array}$$

Step 2. Start with the units column and subtract the subtrahend from the minuend. Since the 7 is larger than the 2, borrow from the tens column $(12 - 7 = 5)$:

$$\begin{array}{r} 812 \\ 1\cancel{9}\cancel{2} \\ -87 \\ \hline 5 \end{array}$$

Step 3. Continue subtracting in the other columns $(8 - 8 = 0)$ and $(1 - 0 = 1)$:

$$\begin{array}{r} 812 \\ 1\cancel{9}\cancel{2} \\ -87 \\ \hline 105 \end{array}$$

Subtraction can be checked by adding the answer and the subtrahend. This sum should equal the minuend.

DIVISION

Division is a simplified process of subtracting the same number a given number of times. For example, if 15 were subtracted from 45, and repeated three times, there would be no remainder.

$$45 - 15 = 30$$
$$30 - 15 = 15$$
$$15 - 15 = 0$$

Another way of explaining division is how many times one number is contained in another number. The expressions used for division are:
Dividend — the number to be divided
Divisor — the number to divide by

BASIC MATHEMATICAL OPERATIONS

Quotient — the answer
Remainder — the extra amound of the quotient that is not less than the divisor and therefore not a whole number.

$$\frac{\text{Dividend}}{\text{Divisor}} = \text{Quotient} + \frac{\text{Remainder}}{\text{Divisor}}$$

$$\text{Divisor} \overline{\smash{)}\text{Dividend}}^{\text{Quotient + Remainder}}$$

The steps for division are:
Step 1. Write the numbers as a division problem.
Step 2. Determine the largest possible number, that when multiplied by the divisor, will not be more than the dividend.
Step 3. Subtract the product of step 2 from the dividend.
Step 4. Bring down the next digit from the dividend.
Step 5. Repeat steps 2 and 3.
Step 6. Continue the process until all the digits of the dividend are used.
Step 7. If the final number resulting from the subtraction is not zero, then the division cannot be done evenly. This final number is the remainder.

Example: If 8236 gpm are needed on the fireground and 14 pumpers are available to supply it, how much water must each pump?

Step 1. Write the numbers as a division problem:

$$14 \overline{\smash{)}8236}$$

Step 2. Determine the largest possible number that, when multiplied by the divisor, will not be more than the dividend ($5 \times 14 = 70$).

$$14 \overline{\smash{)}8236}^{5}$$
$$70$$

Step 3. Subtract the product of step 2 from the dividend ($82 - 70 = 12$).

$$14 \overline{\smash{)}8236}^{5}$$
$$70$$
$$12$$

Step 4. Bring down the next digit from the dividend (3):

$$14 \overline{\smash{)}8236}^{5}$$
$$70$$
$$123$$

Step 5. Determine the largest possible number that, when multiplied by the divisor, will not be more than the dividend ($8 \times 14 = 112$).

```
         58
  14 )8236
       70
       123
       112
```

Step 6. Subtract the product of step 5 from the dividend (123 − 112 = 11).

```
         58
  14 )8236
       70
       123
       112
        11
```

Step 7. Bring down the next digit from the dividend (6):

```
         58
  14 )8236
       70
       123
       112
       116
```

Step 8. Determine the largest possible number that, when multiplied by the divisor, will not be more than the dividend (8 × 14 = 112):

```
        588
  14 )8236
       70
       123
       112
       116
       112
```

Step 9. Subtract the product of step 8 from the dividend (116 − 112 = 4). This difference of 4 is the remainder. Each pumper will be required to supply 588 gpm.

```
        588
  14 )8236
       70
       123
       112
       116
       112
         4
```

Chapter 2

Fractions

In addition to whole numbers, the firefighter must also be able to handle computations involving parts of a whole number—expressed as a fraction.

"The fire apparatus driver/operator shall identify and demonstrate the use of fractions, percentages, and decimal fractions in mathematical calculations as required to solve fire department pumper hydraulic problems."*

If the line and the circle below are divided into four parts, then one part would be 1/4 of the total. Two parts would be 2/4, three parts would be 3/4, and four parts would be 4/4, or the whole amount.

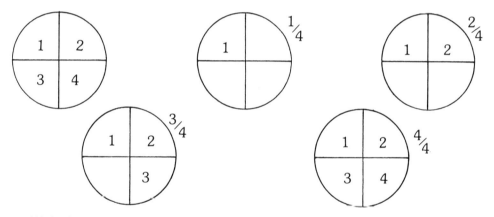

With this explanation then, the fraction line indicates division. The bottom number, called the *denominator*, shows the number of equal parts that the whole is divided into. The *numerator* is the top number and shows how many equal parts of the whole have been taken to make a fraction.

Other terms used to define particular types of fractions are:

Common fraction: A fraction that has both the numerator and denominator expressed as whole numbers.

Proper fraction: A fraction whose numerator is smaller than its denominator.

Improper fraction: A fraction whose numerator is larger than or equal to its denominator.

Complex (compound) fraction: A fraction in which numerator or denominator or both are expressed as a fraction:

Paragraph 304.6. Reprinted with permission from NFPA 1002-1982, Standard for Fire Apparatus Driver/Operator Professional Qualifications, Copyright© 1982, National Fire Protection Association, Quincy, Massachusetts 02269. This reprinted material is not the complete and official position of the NFPA on the referenced subject, which is represented only by the standard in its entirety.

$$\frac{\frac{3}{5}}{\frac{2}{5}} = \frac{3}{5} \div \frac{2}{5}$$

Mixed number: A number which contains a whole number and a fraction.

Reducing common fractions

Fractions can be handled more easily if they are in their lowest terms. This means that they are expressed as simply as possible.

The rule for reducing fractions:

Rule 6: If the numerator and denominator are both multiplied or divided by the same number, the value of the fraction will not change.

Example: Reduce 9/12 to lowest terms.

Step 1. Select the largest number that will divide evenly into both the numerator and denominator. (Answer—3)

Step 2. Divide the numerator and denominator by this number ($9 \div 3 = 3$; $12 \div 3 = 4$).

$$\frac{9 \div 3}{12 \div 3} = \frac{3}{4}$$

Therefore, 3/4 is the fraction expressed in simplest terms.

Reducing improper fractions

Example: Reduce $\frac{28}{6}$ to lowest terms.

Step 1. Divide the numerator by the denominator ($28 \div 6 = 4\frac{4}{6}$):

$$\begin{array}{r} 4 \\ 6\overline{)28} \\ \underline{24} \\ 4 \end{array}$$

Step 2. Reduce the proper fraction to its lowest terms ($4 \div 2 = 2$; $6 \div 2 = 3$):

$$\frac{4 \div 2}{6 \div 2} = \frac{2}{3}$$

Step 3. The fraction can now be expressed as the mixed number 4 2/3.

Sometimes it is necessary to change a mixed number to an improper fraction to perform a mathematical operation.

Example: Change 8 3/4 to an improper fraction.

Step 1. Multiply the whole number by the denominator of the fraction ($8 \times 4 = 32$):

$$8\tfrac{3}{4}$$
$$8 \times 4 = 32$$

Step 2. Add the numerator to the product obtained in step 1 $(32+3=35)$.

$$8 \times 4 + 3 = 35$$

Step 3. Place the result over the denominator. The improper fraction is $\frac{35}{4}$.

$$\frac{8 \times 4 + 3}{4} = \frac{35}{4}$$

Adding fractions

The firefighter will come in contact with drawings and specifications that will contain whole numbers, common fractions and mixed numbers. Here's how these combinations are handled so that they can be added.

Example: A nozzle has tips with the following diameters available: 3/4, 7/8, 1, 1 1/2, 1 1/4 and 1 1/8 inches. What is the sum of the nozzle diameters?

Step 1. To add fractions, all the denominators must be the same. This means finding the smallest number that can be divided by all the denominators, which is 8.

Step 2. Divide this denominator, called the lowest common denominator (LCD), by the denominator of the first fraction 3/4 $(8 \div 4 = 2)$.

Step 3. Multiply the numerator and denominator of the first fraction by the quotient obtained in step 2 $(3 \times 2 = 6; 4 \times 2 = 8)$:

$$\frac{3}{4} \times \frac{2}{2} = \frac{6}{8}$$

Step 4. Those fractions already expressed in terms of the lowest common denominator do not change:

$$\frac{7}{8} \text{ and } \frac{1}{8}$$

Step 5. Continue putting the other functions in terms of the lowest common denominator:

$$\frac{1}{2} \times \frac{4}{4} = \frac{4}{8}$$

Multiply the numerator and denominator by the quotient obtained in step 2:

$$\frac{1}{4} \times \frac{2}{2} = \frac{2}{8}$$

Step 6. Write the fractions and whole numbers in a vertical column:

$$\frac{6}{8}$$

$$\frac{7}{8}$$

$$1$$

$$1\frac{4}{8}$$

$$1\frac{2}{8}$$

$$1\frac{1}{8}$$

Step 7. Add numerators of fractions:

$$6 + 7 + 4 + 2 + 1 = 20$$

Step 8. Place result over lowest common denominator:

$$\frac{20}{8}$$

Step 9. Reduce to lowest possible terms:

$$\frac{20}{8} = 2\frac{4}{8} = 2\frac{1}{2}$$

Step 10. Add column of whole numbers:

$$1 + 1 + 1 + 1 = 4$$

Step 11. Add the sum of the fractions:

$$4 + 2\frac{1}{2} = 6\frac{1}{2}$$

Subtracting fractions

The need for understanding the subtraction of fractions is also necessary for pump operations.

Subtracting proper fractions

Example: Subtract 5/64 from 23/32.

Step 1. Determine the lowest common denominator—64.
Step 2. Write the fractions, in terms of the lowest common denominator, in a vertical column:

$$\frac{23}{32} \times \frac{2}{2} = \frac{46}{64}$$

$$\frac{5}{64} = \frac{5}{64}$$

FRACTIONS

Step 3. Subtract the numerators:

$$46 - 5 = 41$$

Step 4. Place the result over the lowest common denominator:

$$\frac{41}{64}$$

Subtracting mixed numbers from mixed numbers

Example: Subtract 2 13/64 from 5 5/32.

Step 1. Determine the lowest common denominator: 64.
Step 2. Write the fraction, in terms of the lowest common denominator, in a vertical column:

$$5\frac{5}{32} \times \frac{2}{2} = 5\frac{10}{64}$$

$$2\frac{13}{64} = 2\frac{13}{64}$$

Step 3. If the numerator of the number to be subtracted (13) is larger than the numerator of the other fraction (10), then one unit $\left(\frac{64}{64}\right)$ must be borrowed:

$$5\frac{10}{64} = 4 + \frac{64}{64} + \frac{10}{64}$$

$$= 4\frac{74}{64}$$

Step 4. Subtract the numerators of the fractions (74 − 13 = 61) and the whole numbers (4 − 2 = 2).

$$\begin{array}{r} 4\ \frac{74}{64} \\ 2\ \frac{13}{64} \\ \hline 2\ \frac{61}{64} \end{array}$$

Step 5. Place the fraction result over the lowest common denominator:

$$\frac{61}{64}$$

Step 6. The resulting mixed number is the answer:

$$2\ \frac{61}{64}$$

13

Multiplying fractions

As in the muliplication of whole numbers, multiplication of fractions is a simplified process of adding the same fractions given amount of times.

Example: Multiply 2 3/4 by 5 1/3.

Step 1. Change mixed numbers to improper fractions:

$$2\frac{3}{4} = \frac{4 \times 2 + 3}{4} = \frac{11}{4}$$

$$5\frac{1}{3} = \frac{3 \times 5 + 1}{3} = \frac{16}{3}$$

Step 2. Write the problem as a multiplication problem:

$$\frac{11}{4} \times \frac{16}{3}$$

Step 3. Multiply the numerators:

$$11 \times 16 = 176.$$

Step 4. Multiply the denominators:

$$4 \times 3 = 12.$$

Step 5. Write the new fraction:

$$\frac{176}{12}$$

Step 6. Reduce the fractions to the lowest terms:

$$176 \div 12 = 14\frac{8}{12} = 14\frac{2}{3}$$

Rule 6 states that the numerator and the denominator could both be divided by the same number without changing the value of the fraction. Using this rule, the multiplication process can be simplified by using the method called *cancellation*.

Example: Multiply 2 3/4 by 5 1/3.

Step 1. Change mixed numbers to improper fractions:

$$2\frac{3}{4} = \frac{4 \times 2 + 3}{4} = \frac{11}{4}$$

$$5\frac{1}{3} = \frac{3 \times 5 + 1}{3} = \frac{16}{3}$$

Step 2. Write the problem as a multiplication problem:

$$\frac{11}{4} \times \frac{16}{3}$$

Step 3. Select a number that is common to any numerator and any denominator (4).

Step 4. Divide the numerator and denominator by this factor: $4 \div 4 = 1$; $16 \div 4 = 4$:

$$\frac{11}{\cancel{4}_1} \times \frac{\cancel{16}^4}{3}$$

Step 5. Multiply the numerators:

$$11 \times 4 = 44$$

Step 6. Multiply the denominators:

$$1 \times 3 = 3$$

Step 7. Write the new fraction:

$$\frac{44}{3}$$

Step 8. Reduce the fraction to lowest terms:

$$44 \div 3 = 14\frac{2}{3}$$

Dividing fractions

Division of fractions, like division of whole numbers, is the process of determining how many times one number is contained in another.

Dividing proper fractions

Example: Divide 3/8 by 1/4.

Step 1. Write out the division problem:

$$\frac{3}{8} \div \frac{1}{4}$$

Step 2. Invert the divisor:

$$\frac{1}{4} \text{ to } \frac{4}{1}$$

Step 3. Write the problem as a multiplication problem:

$$\frac{3}{8} \times \frac{4}{1}$$

Step 4. Find the number that is common to any numerator and denominator (4 is common to 8 and 4) and divide by this factor:

$$\frac{3}{\cancel{8}_2} \times \frac{\cancel{4}^1}{1}$$

Step 5. Multiply numerators and denominators (3 × 1 = 3; 2 × 1 = 2):

$$\frac{3}{2} \times \frac{1}{1} = \frac{3}{2}$$

Step 6. Reduce the fraction to lowest terms:

$$\frac{3}{2} = 1\frac{1}{2}$$

Dividing mixed numbers

Example: Divide 6 7/8 by 4 7/32.

Step 1. Change the mixed numbers to improper fractions:

$$6\frac{7}{8} = \frac{8 \times 6 + 7}{8} = \frac{55}{8}$$

$$4\frac{7}{32} = \frac{32 \times 4 + 7}{32} = \frac{135}{32}$$

Step 2. Write out the division problem:

$$\frac{55}{8} \div \frac{135}{32}$$

Step 3. Invert the divisor and write the problem as a multiplication problem.

$$\frac{55}{8} \times \frac{32}{135}$$

Step 4. Find the number that is common to any numerator and denominator (8 is common to 8 and 32; 5 is common to 55 and 135) and divide by these factors:

$$\frac{\cancel{55}^{11}}{\cancel{8}_1} \times \frac{\cancel{32}^4}{\cancel{135}_{27}}$$

Step 5. Multiply numerators and denominators (11 × 4 = 44; 27 × 1):

$$\frac{11}{1} \times \frac{4}{27} = \frac{44}{27}$$

Step 6. Reduce the fraction to lowest terms:

$$\frac{44}{27} = 1\frac{17}{27}$$

Decimal fractions

A common fraction indicates division. If this division is actually carried out, the resulting answer is a decimal fraction.

A decimal fraction can be considered as a common fraction with a denominator that is either 10 or a power of 10 (100, 1000, 10,000, etc.). A power of 10 means 10 times itself a specific number of times ($10 \times 10 = 100$; $10 \times 10 \times 10 = 1000$; $10 \times 10 \times 10 \times 10 = 10,000$).

The denominator in a decimal fraction is replaced by a period called a decimal point. Digits written to the left of the decimal point represent whole numbers. Each movement of one digit to the left *increases* the value of the digit by a power of 10. Digits to the right of the decimal point represent decimal fractions and each movement to the right *decreases* the value of the digit by a power of 10. As earlier discussed, the place values to the right of the decimal point are:

MILLIONS	HUNDRED THOUSANDS	TEN THOUSANDS	THOUSANDTHS	HUNDREDS	TENS	UNITS	.	TENTHS	HUNDREDTHS	THOUSANDTHS	TEN THOUSANDTHS	HUNDRED THOUSANDTHS	MILLIONTHS
2,	5	8	9,	6	3	4	.	5	1	4	9	8	7

If the number 333,333 were written, all the digits are the same, but does each digit have the same value? The value of each digit depends on its location within the number. The number can be expressed as:

```
       300 =      3 × 100
        30 =         3 × 10
         3 =             3 × 1
decimal point   .
    3 tenths =             3 × 1/10
3 hundredths =                 3 × 1/100
3 thousandths =                    3 × 1/1000
```

Three hundred thirty-three and three hundred thirty-three thousandths (333 333/1000)

The numerator of the decimal fraction is the digits to the right of the decimal point. The denominator of a decimal fraction is a 1 followed by one zero for each digit in the decimal fraction:

Decimal Fraction	Numerator	No. of Digits	Denominator	Fraction
.8	8	1	10	8/10
.07	7	2	100	7/100
.063	63	3	1000	63/1000
.0520	52	3	1000	52/1000

Note that in the last example the right-hand zero was dropped. This can always be done with zeroes at the right-hand edge of a decimal fraction as they have no value.

The numbers in the preceding table are read as: 8 tenths; 7 hundredths; 63 thousandths; 52 thousandths.

Decimal fractions are added, subtracted, multiplied and divided in the same way as whole numbers, with some extra steps for handling the decimal point.

Adding decimals

The steps for adding decimals are:

Step 1. Write the number one under each other, lining up the decimal point.

Step 2. Add the numbers in the vertical columns and carry over values greater than 9. Since this is a system of 10, carryover across the decimal point is necessary.

Step 3. Insert the decimal point in the answer in the same place as it appears in each number.

Example: If the distances between the centers of gages A, B, C and D were 3.917, 1.086 and .027 inches respectively, what is the distance between A and D?

Step 1. Arrange the numbers vertically and line up the decimal points:

$$3.917$$
$$1.086$$
$$0.027$$

Step 2. Add the numbers and carry over across the decimal point:

①①②
$$3.917$$
$$1.086$$
$$\underline{0.027}$$
$$5.030$$

Step 3. Insert the decimal point in the answer in the same location as it appears in each number:

5.03 (five and three hundredths inches)

Addition should always be checked by adding the digits in reverse order. If the original was added from top to bottom, check your work by adding from bottom to top.

FRACTIONS

Subtracting decimals

The steps for subtraction of decimals are:
Step 1. Write the numbers one under the other, lining up the decimal points. The subtrahend is on the bottom and the minuend is on the top.
Step 2. Start with the right column and subtract the subtrahend from the minuend.
Step 3. If a number in the subtrahend is larger than the number in the same column of the minuend, borrow 10 from the next column to the left.
Step 4. Insert the decimal point in the same location as it appears in each number.

Example: The inner diameter of one hose is found to be 2.416 inches, while the inner diameter of another hose is 1.029 inches. How much larger is the diameter of the first hose over the second hose?

Step 1. Arrange the numbers vertically and line up the decimal point:

$$\begin{array}{r} 2.416 \\ -1.019 \\ \hline \end{array}$$

Step 2. Start with the right-hand column and subtract. If a number in the subtrahend is larger than the number in the minuend, borrow 10 from the next column to the left. Since 9 is larger than 6, borrow 10 from the hundredths column (16 − 9 = 7).

$$\begin{array}{r} \overset{\circ}{2}.\,4\,\overset{\circ}{\cancel{1}}\,\overset{16}{\cancel{6}} \\ -1.\,0\,2\,9 \\ \hline 7 \end{array}$$

Step 3. Continue subtracting in the other columns, borrowing where necessary:

$$\begin{array}{r} \overset{3}{\cancel{2}}.\,\overset{\overset{10}{\circ}}{\cancel{4}}\,\overset{\circ}{\cancel{1}}\,\overset{16}{\cancel{6}} \\ -1.\,0\,2\,9 \\ \hline 1.\,3\,8\,7 \end{array}$$

Step 4. Insert the decimal point in the answer in the same location as it appears in each number:

1.387 inches (one and three hundred eighty-seven thousandths inches)

Subtraction can be checked by adding the answer and the subtrahend, which should equal the minuend.

Multiplying decimals

The steps for multiplying decimals are:
Step 1. Place the multiplier below the multiplicand, with the right-hand digits lined up.
Step 2. Multiply in the same manner as used for whole numbers.
Step 3. Count the number of decimal places (numbers to the right of the

decimal point) in the multiplier and the multiplicand.

Step 4. Place the decimal point in the answer by starting at the right-hand digit and counting off the number of places from step 3.

Example: If 1 square inch of water 1 foot high weighs .434 pounds, how much will 1 square inch of water 7.26 feet high weigh?

Step 1. Place the number with the right-hand digits lined up:

$$7.26$$
$$.4\ 34$$

Step 2. Multiply in the same manner as used for whole numbers:

```
        ① ②
         ①
        ① ②
       7.2 6
      .4 3 4
      ───────
       2.9 0 4
       2 1 7 8
     2 9 0 4
     ─────────
     3 1 5 0 8 4
```

Step 3. Count off the number of decimal places (5).
Step 4. Place the decimal point in the answer by starting at the right-hand edge and counting the number from step 3 (5):

$$3.15084$$

Step 5. Write the answer:

3.15084 pounds

Multiplication can be checked by reversing the multiplier and the multiplicand. Always check the location of the decimal point.

Multiplying decimals by powers of 10

The distributive rule of multiplication (rule 5) permits the development of an easy method of multiplying decimal fractions by powers of 10. This method says that for each zero in the multiplier, move the decimal point one place to the right.

Example: Multiply .434 × 100.

Step 1. Determine the number of zeros in the multiplier 100 (2).
Step 2. Move the decimal point two places to the right:

43.4

Step 3. Write the answer:

.434 × 100 = 43.4

FRACTIONS

Dividing decimals

The steps for dividing decimals are:

Step 1. Write the figures as a division problem, in the same way as for whole numbers.

Step 2. Move the decimal point of the divisor to the extreme right and count the number of places moved. This makes the divisor a whole number.

Step 3. Move the decimal point of the dividend the same number of places determined in step 2. If there are not enough digits in the dividend, add zeros before placing the decimal point.

Step 4. Place the decimal point in the quotient above the spot determined in step 3.

Step 5. Divide in the same manner used for whole numbers.

Step 6. Continue the division until the quotient has the number of decimal places needed for the answer. Add zeroes to the right of the decimal point if more places are needed in the dividend.

Example: Divide 16.73 by 2.304, use three decimal places in the answer.

Step 1. Write the figures as a division problem:

$$2.304 \overline{)16.73}$$

Step 2. Move the decimal point of the divisor to the extreme right three places:

$$2304.$$

Step 3. Move the decimal point of the dividend three places to the right:

$$16730.$$

Step 4. Place the decimal point in the quotient:

$$2304 \overline{)16730.}$$

Step 5. Divide in the same manner used for whole numbers. Add three zeroes to the right of the decimal point so that the answer will have three decimal places:

```
              7.261
      2304 )16730.000
            16128
             602 0
             460 8
             141 20
             138 24
               2 960
               2 304
                 656
```

Step 6. Write the answer:

$$16.73 \div 2.304 = 7.261$$

Dividing decimals by powers of 10

As in the case of multiplication, the distributive rule permits the development of an easy method of dividing decimal fractions by powers of 10. This method says that for each zero in the divisor, move the decimal point one place to the left.

Example: Divide 2.304 by 1000.

Step 1. Determine the number of zeroes in the divisor, 1000 (3).
Step 2. Move the decimal point three places to the left:

.002304

Step 3. Write the answer:

.002304

Rounding off decimals

It is sometimes necessary to delete some of the least significant digits in a decimal fraction. The number of places in the answer is determined by the accuracy required. The steps for rounding off are:
Step 1. Determine the number of places required.
Step 2. Write the number using one more place than that determined in step 1.
Step 3. If the right-hand digit is 6 or greater, increase the next digit to the left by 1.
Step 4. If the right-hand digit is 4 or less, leave the next digit to the left as it is.
Step 5. If the right-hand digit is 5 and it is preceded by an odd number, increase the digit by 1.
Step 6. If the right-hand digit is 5 and it is preceded by an even number, leave the next digit to the left as it is.

Example: Round off 3.72024 to three decimal places; .9876981 to two decimal places; 1.064566 to three decimal places.

Step 1. Determine the number of places required:

3.72024   .9876981   1.064566
  (3)   (2)   (3)

Step 2. Write the number using one more place than that determined in step 1:

3.7202   .987   1.0645

Step 3. Round the numbers according to the rules:

3.720   .99   1.064

Decimal equivalents

Many times when working with numerous fractions, especially in measurements of length, it is easier to handle the problem as a decimal fraction. To make the conversion from common fraction to decimal fraction easier, a standard chart (table 1) for common measurement fractions has been developed.

TABLE 1 Decimal Equivalent

1/64	.015625	17/64	.265625	33/64	.515625	49/64	.765625
1/32	.03125	9/32	.28125	17/32	.53125	25/32	.78125
3/64	.046875	19/64	.296875	35/64	.546875	51/64	.796875
1/16	.0625	5/16	.3125	9/16	.5625	13/16	.8125
5/64	.078125	21/64	.328125	37/64	.578125	53/64	.828125
3/32	.09375	11/32	.34375	19/32	.59375	27/32	.84375
7/64	.109375	23/64	.359375	39/64	.609375	55/64	.859375
1/8	.125	3/8	.375	5/8	.625	7/8	.875
9/64	.140625	25/64	.390625	41/64	.640625	57/64	.890625
5/32	.15625	13/32	.40625	21/32	.65625	29/32	.90625
11/64	.171875	27/64	.421875	43/64	.671875	59/64	.921875
3/16	.1875	7/16	.4375	11/16	.6875	15/16	.9375
13/64	.203125	29/64	.453125	45/64	.703125	61/64	.953125
7/32	.21875	15/32	.46875	23/32	.71875	31/32	.96875
15/64	.234375	31/64	.484375	47/64	.734375	63/64	.984375
1/4	.25	1/2	.5	3/4	.75	1	1.

The decimal equivalent can be read directly from the chart. Conversely, if the decimal equivalent is known, the chart will provide the nearest fraction equivalent.

Chapter 3

Algebra

In addition to the basics of mathematics, "The fire apparatus driver/operator shall demonstrate the use of simple algebraic formulas required to solve fire department pumper hydraulic problems."*

Algebra is an extension of the basic mathematical procedures. However, instead of using numbers, algebra replaces the numbers with letters. These letters can be used to represent a general number. By using these general numbers, mathematical definitions can be developed that will apply any time, no matter what numbers are used. The mathematical rules continue to apply, with certain additional rules for the handling of algebraic symbols.

In Chapter 1, some mathematical symbols were defined. These symbols [+, −, ×, ÷, =, ()] represent the same operations in algebra, but additional rules and procedures are also necessary. These rules deal with the mathematical designation of groups, subscript notations, and the performance of algebraic procedures.

Some additional definitions required for discussing algebraic operations are given below. The examples are based upon the expression $3ax + 4y - z$.

Expression	Explanation	Example
Algebraic expression	A group of symbols that represent a number.	$3ax + 4y - z$
Term	A combination of symbols between + (plus) or − (minus) signs.	$+3ax$, or $+4y$, or $-z$ [Note that when no sign is included in an algebraic expression, a plus sign is understood ($+3ax$); and if a letter is present without a number (z), the number 1 is understood.]
Factors	The individual parts of the terms.	3, a, and x are each factors of the term $3ax$.
Numerical Coefficient	The number in a term.	3 is the numerical coefficient of ax.

---

*Paragraph 3-4.7. Reprinted with permission from NFPA 1002-1982, Standard for Fire Apparatus Driver/Operator Professional Qualifications, Copyright© 1982, National Fire Protection Association, Quincy, Massachusetts 02269. This reprinted material is not the complete and official position of the NFPA on the referenced subject, which is represented only by the standard in its entirety.

Letters and subscripts

In algebra, letters are used to represent numbers. Sometimes a specific letter is used to designate values for a specific quantity. For example, when discussing fire service hydraulics, p is usually used to represent pressure in pounds per square inch, while d stands for the diameter of a nozzle in inches.

Letters, then, are used to express formulas—general expressions that state a fact. These general expressions are true no matter what numerical values are used in the formula to replace the letters. For example, if A = area of a rectangle, l = length of a rectangle, and w = width of a rectangle, then the formula A = w × l, can be stated. This formula would be true for any value of width and length. It is important to distinguish between capital and small letters because the same letter can be used as a capital and as a small letter in the same formula and mean different quantities.

Another way of distinguishing different values of the same quantity is by the use of subscripts. The subscript number has no numerical value. For example, if three successive pressure measurements were made, they might be designated as $p_1 = 58$ psi, $p_2 = 63$ psi, and $p_3 = 59$ psi. The subscripts 1, 2 and 3 just show a sequence of measurement.

In certain cases, a particular letter is chosen to represent a constant value that does not vary from formula to formula. Such a symbol is the Greek letter pi (π) that represents the value 3.14159. The formula for the area (A) of a circle is $A = \pi r^2$, with r signifying the radius of a circle. The circumference (C) of a circle is $C = 2\pi r$. In both of these formulas, the symbol π has the same value, 3.14159. In working with a constant in a formula, it can be handled mathematically as any number. However, it is easier to handle π than 3.14159 and the substitution of the number for the symbol can be made after the necessary manipulation of the formula is completed.

Adding symbols

To add terms, which are made up of individual factors, all factors except the numerical coefficient must be the same. The steps for addition are:

Step 1. Group the factors that are alike together.
Step 2. Add the numerical coefficients of the like factors.
Step 3. Rewrite the final answer.

Example: Add 3xy + 2z + 5yz + 7xy + 3y + 4xz + z + 2xy + yz + 4y.

Step 1. Group like factors together:

$$3xy + 7xy + 2xy$$
$$2z + z$$
$$5yz + yz$$
$$3y + 4y$$
$$4xz$$

Step 2. Add the numerical coefficients of the like factors:

$$3xy + 7xy + 2xy = 12xy$$
$$2z + z = 3z$$
$$5yz + yz = 6yz$$
$$3y + 4y = 7y$$
$$4xz = 4xz$$

Step 3. Rewrite the final answer:

$$12xy + 3z + 6yz + 7y + 4xz$$

Subtracting symbols

As in addition, subtraction of symbols requires that all factors except the numerical coefficient within the term be the same. The steps for subtraction are:

Step 1. Group the factors that are alike together.
Step 2. Perform the mathematical operations, addition and/or subtraction, for the numerical coefficients of the like factors.
Step 3. Rewrite the answer.

The rules for handling addition and subtraction can be stated as follows:

Rule 7: When combining factors with the same sign (either plus or minus) add the numerical coefficients and place the sign in front of the numerical coefficient of the answer.

Rule 8. When combining factors with opposite signs, subtract the numerical coefficients and place the sign of the larger coefficient in front of the numerical coefficient of the answer.

Example: ab + 6a – 5c – 2ab – 2a + 2b + 6c – 4bc – 4a – 3c – 2bc.

Step 1. Group the factors that are alike together:

$$+ ab - 2ab$$
$$+ 6a - 2a - 4a$$
$$- 5c + 6c - 3c$$
$$+ 2b$$
$$- 4bc - 2bc$$

Step 2. Perform the mathematical operations indicated:

$$+ ab - 2ab = - ab$$
$$+ 6a - 2a - 4a = 0$$
$$- 5c + 6c - 3c = - 2c$$
$$+ 2b = + 2b$$
$$- 4bc - 2bc = - 6bc$$

Step 3. Rewrite the final answer:

$$- ab - 2c + 2b - 6bc$$

Subtraction of negative numbers requires the use of an additional rule:

Rule 9: With negative numbers used in subtraction, change the sign of subtrahend and add.

Example: Subtract 4x – (–2x)

Step 1. Using rule 9, change the sign of the factor (–2x) to +2x.
Step 2. Add the factors: 4x + (+2x) = 6x.

Multiplying symbols

Multiplication of symbols can be accomplished whether or not the terms or factors are the same. The steps for multiplication are:
Step 1. Multiply the numerical coefficients.
Step 2. Multiply the symbols of the terms.

ALGEBRA

Step 3. Determine the sign by using the rules:

$$+ \times + = +$$
$$+ \times - = -$$
$$- \times - = +$$

Step 4. Combine the numerical coefficients, the symbols, and the sign.

Examples: (3x) (4y) (2z); (−2a) (4b); (−6c) (−3d)

Step 1. Multiply the numerical coefficients:

Example 1	**Example 2**	**Example 3**
(3)(4)(2) = 24	(2)(4) = 8	(6)(3) = 18

Step 2. Multiply the symbols:

Example 1	**Example 2**	**Example 3**
(x)(y)(z) = xyz	(a)(b) = ab	(c)(d) = cd

Step 3. Determine the sign:

Example 1	**Example 2**	**Example 3**
(+)(+)(+) = +	(−)(+) = −	(−)(−) = +

Step 4. Combine:

Example 1	**Example 2**	**Example 3**
+24xyz	−8ab	+18cd

There are various methods for indicating multiplication in algebra. It is necessary to be familiar with all the methods.

1. a(b) = ab
2. (a)(b) = ab
3. a × b = ab
4. a · b = ab

Dividing symbols

Division of terms with symbols is very similar to arithmetical division. The steps for division are:
Step 1. Write the problem in the form of a fraction.
Step 2. Divide the numerical coefficients of the numerator and denominator.
Step 3. Cancel the symbols that are common to the terms in both the numerator and denominator.
Step 4. Determine the sign by using the rules:

$$+ \div + = +$$
$$+ \div - = -$$
$$- \div - = +$$

Step 5. Write the final answer.

Examples: 36xy ÷ 12y; −4a ÷ 3abc; −5de ÷ −8ef

Step 1. Write the problem as a fraction:

Example 1	Example 2	Example 3
$\dfrac{36xy}{12y}$	$\dfrac{-4a}{3abc}$	$\dfrac{-5de}{-8ef}$

Step 2. Divide the numerical coefficients of the numerator and denominator:

$$\dfrac{36}{12} = 3 \qquad \dfrac{4}{3} = 1.33 \qquad \dfrac{5}{8} = .625$$

Step 3. Divide the symbols:

$$\dfrac{x\cancel{y}}{\cancel{y}} = x \qquad \dfrac{\cancel{a}}{\cancel{a}bc} = \dfrac{1}{bc} \qquad \dfrac{d\cancel{e}}{\cancel{e}f} = \dfrac{d}{f}$$

Step 4. Determine the sign:

$$\dfrac{+}{+} = + \qquad \dfrac{-}{+} = - \qquad \dfrac{-}{-} = +$$

Step 5. Write the final answer:

$$+3x \qquad \dfrac{-1.33}{bc} \qquad \dfrac{.625d}{f}$$

Algebraic grouping

In order to avoid misinterpretation, a series of symbols has been developed for indicating the sequence of steps necessary to perform the mathematical operations. These grouping symbols are: () parentheses; [] brackets; { } braces. Rules for handling the grouping symbols are:

Rule 10. If a plus sign (+) precedes a grouping symbol, perform the operations within the symbol exactly as indicated.

Rule 11. If a minus sign (−) precedes a grouping symbol, change the sign within the grouping symbol (a minus to a plus or a plus to a minus) and then change the sign preceding the grouping symbol to a plus. Perform the operations within the symbol.

Rule 12. If a coefficient precedes the grouping symbol, each term within the symbol must be multiplied by the coefficient.

Example: $6\{5[4(3xy + 2y) - 6(xy - 2z)] + 7y\} - 6z$. Find the value of this algebraic expression if $x=2$, $y=3$ and $z=4$.

Step 1. Perform the first operation within the parentheses, using rules 10 and 12 for + sign grouping:

$$4(3xy + 2y) = 12xy + 8y$$

Step 2. Perform the second operation within the parentheses, using rules 11 and 12 for + sign grouping:

$-6(xy - 2z)$
Change $+xy$ to $-xy$
Change $-2z$ to $+2z$
Change -6 to $+6$
Multiply by the coefficient $6(-xy + 2z) = -6xy + 12z$

Step 3. Rewrite the problem with the work performed in steps 1 and 2:

$$6\{5[12xy + 8y - 6xy + 12z] + 7y\} - 6z$$

Step 4. Combine similar terms within the brackets:

$$+12xy - 6xy = +6xy$$

Step 5. Rewrite the problem:

$$6\{5[6xy + 8y + 12z] + 7y\} - 6z$$

Step 6. Perform the operation within the brackets, using Rules 10 and 12 for + sign grouping:

$$5[6xy + 8y + 12z]$$
$$30xy + 40y + 60z$$

Step 7. Rewrite the problem:

$$6\{30xy + 40y + 60z + 7y\} - 6z$$

Step 8. Combine similar terms within the braces:

$$40y + 7y = 47y$$

Step 9. Rewrite the problem:

$$6\{30xy + 47y + 60z\} - 6z$$

Step 10. Perform the operation within the braces, using rules 10 and 12 for + sign grouping:

$$180xy + 282y + 360z - 6z$$

Step 11. Combine similar terms:

$$360z - 6z = 354z$$

Step 12. Rewrite the problem:

$$180xy + 282y + 354z$$

Step 13. Substitute the numerical values $x=2$, $y=3$, $z=4$:

$$180(2)(3) + 282(3) + 354(4)$$
$$1080 + 846 + 1416 = 3342$$

Equations

As explained in the introduction to this unit, formulas are general expressions that state a fact. An equation is a statement that expresses an equality and indicates that the quantity on one side of the = sign is the same as that on the other side. Whenever a formula is expressed, an equality is also expressed. However, not all equations can be considered formulas.

A formula is said to have a left-hand and a right-hand side, separated by an equal sign. Equations can be solved by applying certain basic rules:

Rule 13: The value of an equation is not changed if:
1. The same value is added to both sides of the equation.
2. The same value is subtracted from both sides of the equation.

3. Both sides of the equation are multiplied by the same number.
4. Both sides of the equation are divided by the same number.
5. A factor is moved from one side of the equation to the other side and its sign is changed. This is called transposing.

Example: Solve the equation x − 4 = 11.

Step 1. In order to solve the equation, the unknown quantity, x, must be isolated. This can be accomplished by using rule 13(1) and adding 4 to each side of the equation.

$$x - 4 = 11$$
$$+ 4 \quad +4$$

Step 2. Add both sides of the equation:

$$\begin{aligned} x - 4 &= 11 \\ + 4 &\quad +4 \\ \hline x + 0 &= 15 \\ x &= 15 \end{aligned}$$

Step 3. Check the answer by substituting the answer in the original equation:

$$15 - 4 = 11$$
$$11 = 11$$

Example: Solve the equation a + 3a + 6 = −10.

Step 1. In order to solve the equation, the unknown quantity, a, must be isolated. This can be accomplished by combining a and 3a and using rule 13(2). Subtract 6 from each side of the equation.

$$\begin{aligned} a + 3a + 6 &= -10 \\ 4a + 6 &= -10 \\ -6 &\quad -6 \end{aligned}$$

Step 2. Subtract both sides of the equation:

$$\begin{aligned} 4a + 6 &= -10 \\ -6 &\quad -6 \\ \hline 4a + 0 &= -16 \\ 4a &= -16 \end{aligned}$$

Step 3. In order to isolate a, both sides of the equation must be divided by 4, using rule 13(4).

$$\frac{4a}{4} = -\frac{16}{4}$$

Step 4. Divide both sides of the equation:

$$\frac{\cancel{4}a}{\cancel{4}} = -\frac{\cancel{16}^{4}}{\cancel{4}}$$
$$a = -4$$

ALGEBRA

Step 5. Check the answer by substituting the answer in the original equation:

$$-4 + 3(-4) + 6 = -10$$
$$-4 - 12 + 6 = -10$$
$$-10 = -10$$

Example: Solve the equation $\dfrac{c}{6} - 4 = 5$.

Step 1. In order to solve the equation, the unknown quantity c must be isolated. This can be accomplished by using rule 12(1) and adding 4 to both sides of the equation.

$$\dfrac{c}{6} - 4 = 5$$
$$+4 \quad +4$$

Step 2. Add both sides of the equation:

$$\dfrac{c}{6} + 0 = 9$$

Step 3. In order to isolate c, both sides of the equation must be multiplied by 6, using Rule 13(3):

$$\dfrac{c}{6} \times 6 = 9(6)$$

Step 4. Multiply both sides of the equation:

$$\dfrac{c}{\cancel{6}} \times \cancel{6}^{1} = 54$$

$$c = 54$$

Step 5. Check the answer by substituting the answer in the original equation:

$$\dfrac{54}{6} - 4 = 5$$

$$\dfrac{\cancel{54}^{9}}{\cancel{6}_{1}} - 4 = 5$$

$$9 - 4 = 5$$
$$5 = 5$$

Example: Solve the equation $6a + 7 - 2a = -3a - 14$.

Step 1. In order to solve the equation, the unknown quantity, a, must be isolated. This can be accomplished by using Rules 13(1) and 13(5), transposing.

$$6a + 7 - 2a = -3a - 14$$
$$4a + 7 = -3a - 14$$
$$+3a$$
$$4a + 7 = -3a - 14$$
$$-7$$

Step 2. Combine terms:

$$4a + 3a = -14 - 7$$
$$7a = -21$$

Step 3. Divide both sides of the equation by 7, using Rule 13(4):

$$\frac{\cancel{7}a}{\cancel{7}} = \frac{\cancel{21}^{3}}{\cancel{7}_{1}}$$

$$a = -3$$

Step 4. Check the answer by substituting the answer in the original equation:

$$6(-3) + 7 - 2(-3) = -3(-3) - 14$$
$$-18 + 7 + 6 = +9 - 14$$
$$-5 = -5$$

Chapter 4

Ratio, Proportion, Percentage, Powers and Roots

The last area of mathematics that potential pump operators need to review is ratio, proportion, percentage, powers and square roots.

"The fire apparatus driver/operator shall demonstrate the use of proportions in mathematical calculations as required to solve fire department pumper hydraulics problems.

"The fire apparatus driver/operator shall identify and demonstrate the determination and use of square roots at required to solve fire department pumper hydraulic problems."*

RATIO

It is often necessary to make a comparison of like quantities. This can be done by making a ratio, that can be expressed simply as a fraction. For example, pumper 1 which is 12 years old can be compared to pumper 2 which is six years old by expressing the ratio:

$$\frac{\text{Pumper 1 age}}{\text{Pumper 2 age}} = \frac{12 \text{ years}}{6 \text{ years}} = \frac{2}{1}$$

This means that pumper 1 is twice as old as pumper 2. Since a ratio can be expressed as a fraction, it can be reduced to lowest terms without changing in value. It is important to remember that ratios can be used to compare like quantities. This leads to rule 14 which states:

Rule 14: In a ratio, both the numerator and denominator must be expressed in the same units.

In addition to being expressed as a fraction or a division problem, a ratio can be shown in the form 6 : 12. This form is read as "6 is to 12," with 6 as the numerator and 12 as the denominator. The steps for determining a ratio are:

Step 1. Determine the quantities that are the same.
Step 2. Write the quantities as a ratio.
Step 3. Reduce the ratio to lowest terms.

Paragraphs 3-4.4 and 3-4.5. Reprinted with permission from NFPA 1002-1982, Standard for Fire Apparatus Driver/Operator Professional Qualifications. Copyright© 1982, National Fire Protection Association, Quincy, Massachusetts 02269. This reprinted material is not the complete and official position of the NFPA on the referenced subject, which is represented only by the standard in its entirety.

Example: The second floor of a building is 125 feet wide by 175 feet long. What is the ratio of the width to the length?

Step 1. Determine the quantities that are the same:

$$125 \text{ feet}$$
$$175 \text{ feet}$$

Step 2. Write the quantities as a ratio:

$$125 : 175$$

$$\frac{125}{175}$$

Step 3. Reduce the ratio to lowest terms (125 ÷ 25 = 5; 175 ÷ 7 = 7):

$$\frac{\cancel{125}^{\,5}}{\cancel{175}_{\,7}} = \frac{5}{7}$$

The width is 5/7 of the length.

PROPORTION

A proportion is the equality of two ratios. It means that the ratio of a quantity is equal to the ratio of another quantity. For example, 5/10 = 10/20 is a proportion showing the equality of the two ratios.

Another way of writing a proportion is:

$$5 : 10 :: 10 : 20$$

This can be read as "5 is to 10 as 10 is to 20." The symbol : is read as "is to" and the symbol :: is read as "as." When written this way, the end or outside terms are called the *extremes*, while the middle or inside terms are called the *means*.

Rule 15: For proportions, the product of the mean equals the product of the extremes.

As an example, the general proportion is

$$a : b :: c : d, \text{ or}$$
$$\frac{a}{b} = \frac{c}{d}$$

Using rule 15, a and d are the extremes and b and c are the means. Therefore, ad = bc. This rule is used to solve a proportion when one of the terms is unknown.

In ratios, only like quantities can be compared. In proportions, however, each ratio can contain unlike quantities, provided that when setting up the proportion, the unlike terms are compared in the same order. This results in the development of two different types of proportions, the direct and indirect.

A *direct proportion* is a proportion in which the unlike quantities change in the same order in each ratio. An *indirect* or *inverse proportion* is one in which one ratio gets larger while the other ratio gets smaller.

RATIO, PROPORTION, PERCENTAGE, POWERS AND ROOTS

The proportion

$$\frac{60 \text{ miles}}{120 \text{ miles}} = \frac{4 \text{ minutes}}{8 \text{ minutes}} \text{ or } 60 : 120 :: 4 : 8$$

is a direct proportion because both ratios increase in the same direction (as the miles increase so does the time required).

In a proportion dealing with the number of people needed to do a job, the more individuals involved, the less time that is necessary. If it takes six people 12 minutes to do a job, then it will take 24 people 3 minutes to do the same job. The proportions can be set up as follows:

$$\frac{6 \text{ people}}{24 \text{ people}} = \frac{3 \text{ minutes}}{12 \text{ minutes}} \text{ or } 6 : 24 :: 3 : 12$$

This is an indirect proportion because as one quantity increases, the other quantity decreases.

The steps for solving proportion problems are:

Step 1. Determine the three known values and assign a letter to the unknown value that is to be computed.

Step 2. Determine whether the proportion has a direct or indirect relationship.

Step 3. Write the ratios as a proportion.

Step 4. Solve the proportion using rule 15.

Step 5. Check the answer by substituting the computed value in the original problem and reworking.

Example: An elevated water supply tank holds 10,000 gallons of water when filled to a depth of 50 feet. How many gallons will it contain when the depth is 29 feet?

Step 1. Determine the three known values and the unknown:

10,000 gallons
G = unknown gallons
50 feet
29 feet

Step 2. Determine whether direct or indirect: direct (the amount of water in the tank *decreases* as the water depth *decreases*.

Step 3. Write the ratio as a proportion:

10,000 : G :: 50 : 29

$$\frac{10,000}{G} = \frac{50}{29}$$

Step 4. Solve the proportion using rule 15.

⌒Mean⌒
10,000 : G :: 50 : 29
⌣Extreme⌣

50 G = 29,000
G = 58,000 gallons

Step 5. Check the answer:

$$10{,}000\ (29) = 50\ (5800)$$
$$290{,}000 = 290{,}000$$

Example: If it takes seven firefighters 28 minutes to perform a particular evolution, how many firefighters are needed to do the job in four minutes?

Step 1. Determine the unknown values and the unknown:

$$\begin{aligned}&7\text{ firefighters}\\ F = {}&\text{unknown firefighters}\\ &28\text{ minutes}\\ &4\text{ minutes}\end{aligned}$$

Step 2. Determine whether direct or indirect. This is indirect (the number of firefighters needed to do the job *increases* when the time necessary *decreases*.)

Step 3. Write the ratio as a proportion:

$$7 : F :: 4 : 28$$

$$\frac{7}{F} = \frac{4}{28}$$

Step 4. Solve the proportion using rule 15:

$$4F = 196$$
$$F = 49$$

Step 5. Check the answer:

$$28\ (7) = 4\ (49)$$
$$196 = 196$$

PERCENTAGES

The word *percent* is another way of saying hundredths and is indicated by a % sign. The word *percentage* means dividing something into 100 equal parts. There are three ways of expressing hundredths:

$$\frac{1}{100}\ \text{common fraction}$$

.01 decimal fraction

1% percent

The steps for finding a percent of a number are:
Step 1. Change the percent to decimal form.
Step 2. Multiply the number by the decimal.

Example: What is 17% of 1264 gallons?

Step 1. Change the percent to a decimal:

$$17\% = .17$$

Step 2. Multiply the number by the decimal:

$$\begin{array}{r} 1264 \\ \underline{.17} \\ 8848 \\ \underline{1264} \\ 214.88 \end{array}$$

214.88 gallons

To calculate percentage, it is necessary to know the rate of percent and the amount with which the comparison is to be made. The formula that states this is: p = rb, where p = percentage, r = rate (%), b = base (quantity with which the comparison is to be made). Using algebra, this formula can be rewritten as:

$$\frac{p}{r} = b \text{ and } \frac{p}{b} = r$$

The particular form of the formula used depends on which of the quantities are known.

Example: A fire department employs 73 firefighters. If an 11 percent increase is authorized, how many firefighters will be hired and what will be the total employment?

Step 1. Write the formula:

$$p = rb$$
$$r = .11;\ b = 73$$

Step 2. Substitute in the formula and perform the mathematical operation:

$$p = .11\ (73)$$
$$p = 8.03 \text{ firefighters}$$
$$\text{total department} = 73 + 8$$
$$= 81 \text{ firefighters}$$

Example: A particular engine in a pumper is rated to develop 320 hp. When actually measured, the engine develops 249.6 hp. What percent efficient is the engine?

Step 1. Write the formula:

$$r = \frac{p}{b}$$

$$p = 249.6;\ b = 320$$

Step 2. Substitute in the formula and perform the mathematical operation:

$$r = \frac{249.6}{320}$$

$$r = .78$$

$$r = 78\%$$

Example: An engine in a pumper is rated as being 82 percent efficient and actually develops 270 hp. What is the theoretical horsepower of the engine?

Step 1. Write the formula:

$$b = \frac{p}{r}$$

$$p = 270; r = .82$$

Step 2. Substitute in the formula and perform the mathematical operation:

$$b = \frac{270}{.82}$$

$$b = 329.27 \text{ hp}$$

POWERS

As previously discussed, an easier way to express the sum $5+5+5$ is to write 5×3. There is also a shorthand notation for repeated multiplication. For example, 5×5 can be written as 5^2 and $5 \times 5 \times 5$ can be written as 5^3. The number 5 is called the *base;* the numbers 2 and 3 are called the *exponents;* and the product of 5^2 or 5^3 is called the *power* of the factors.

The units of measurement are very important when considering powers. A unit of measure, when multiplied by itself, is a square (inches times inches equals square inches). At the conclusion of the mathematical operations, the correct units must be calculated.

The rules for handling exponents are:

Rule 16: The basic rules of exponents are:

A. $(x^a)(x^b) = x^{a+b}$

B. $(x^a)^b = x^{ab}$

C. $(xy)^a = (x^a)(y^a)$

D. $\dfrac{x^a}{x^b} = x^{a-b}$

E. $\left(\dfrac{x}{y}\right)^a = \dfrac{x^a}{y^a}$

F. $x^0 = 1$

The steps for performing operations with powers are:
Step 1. Write the problem.
Step 2. Determine the correct rule to use and perform the mathematical operation.
Step 3. Calculate the correct units.

Example: Calculate $(3^2)(3^3)$; $(3^2)^3$; $(3 \cdot 4)^3$; $\dfrac{3^5}{3^2}$; $\left(\dfrac{3}{4}\right)^2$

Step 1. Write the problem:

Example 1	Example 2	Example 3	Example 4	Example 5
$(3^2)(3^3)$	$(3^2)^3$	$(3 \cdot 4)^3$	$\dfrac{3^5}{3^2}$	$\left(\dfrac{3}{4}\right)^2$

RATIO, PROPORTION, PERCENTAGE, POWERS AND ROOTS

Step 2. Determine the correct rule to use and perform the mathematical operation:

Rule 16A	Rule 16B	Rule 16C	Rule 16D	Rule 16E
3^{2+3}	$3^{2 \times 3}$	$3^3 \times 4^3$	3^{5-2}	$\dfrac{3^2}{4^2}$
3^5	3^6	27×64	3^3	
243	729	1728	27	$\dfrac{9}{16}$

Step 3. Calculate the correct units

fifth power	sixth power	third power	third power	square

Example: If a sphere has a diameter of 3½ inches, what is its volume using the formula $v = \pi D^3 \div 6$, where $\pi = 3.14$; D = 3.5 inches, and $D^3 = (3.5)^3$ cubic inches?

Step 1. Write the problem:

$$v = \frac{\pi D^3}{6}$$
$$\pi = 3.14$$
$$D = 3.5 \text{ inches}$$
$$D^3 = (3.5)^3 \text{ cubic inches}$$

Step 2. Perform the mathematical operations:

$$v = \frac{3.14 \times 3.5 \times 3.5 \times 3.5}{6}$$
$$v = \frac{134.627}{6}$$
$$v = 22.44$$

Step 3. Calculate the correct units:

$$\text{inches} \times \text{inches} \times \text{inches} = \text{inches}^3$$
$$= \text{cubic inches}$$
$$v = 22.44 \text{ cubic inches}$$

ROOTS

A root is the opposite of a power. It says that a number, when multiplied by itself, will give the original number. For example, the square root of 25 asks what number, when multiplied by itself, will give 25. Similarly, the cube root of 27 asks what number when multiplied by itself three times will give 27.

The mathematical shorthand method of writing square root is the radical sign $\sqrt{}$. To indicate a cube root, the radical sign is written $\sqrt[3]{}$.

Very few square roots work out to a whole number. The square root of 9 is exactly 3, but the square root of 10 is not a whole number. The square root of 10 can be rounded off to 3.16. For this reason it is necessary to calculate most of the square roots. However, it is not the purpose of this text to teach the

PUMP OPERATORS HANDBOOK

mathematical method for determining square root. Instead, Table 1 has been provided to give the square root of numbers from 1 to 100, to two decimal places.

TABLE 1. Square Roots

No.	Sq. Root	No.	Sq. Root	No.	Sq. Root	No.	Sq. Root
1	1.00	26	5.10	51	7.14	76	8.72
2	1.4	27	5.20	52	7.21	77	8.78
3	1.73	28	5.29	53	7.28	78	8.83
4	2.00	29	5.38	54	7.35	79	8.89
5	2.24	30	5.48	55	7.42	80	8.94
6	2.45	31	5.57	56	7.48	81	9.00
7	2.65	32	5.66	57	7.55	82	9.06
8	2.83	33	5.74	58	7.62	83	9.11
9	3.00	34	5.83	59	7.68	84	9.16
10	3.16	35	5.92	60	7.75	85	9.22
11	3.32	36	6.00	61	7.81	86	9.27
12	3.46	37	6.08	62	7.87	87	9.33
13	3.61	38	6.16	63	7.94	88	9.38
14	3.74	39	6.24	64	8.00	89	9.43
15	3.87	40	6.32	65	8.06	90	9.49
16	4.00	41	6.40	66	8.12	91	9.54
17	4.12	42	6.48	67	8.18	92	9.59
18	4.24	43	6.56	68	8.25	93	9.64
19	4.36	44	6.63	69	8.31	94	9.70
20	4.47	45	6.71	70	8.37	95	9.75
21	4.58	46	6.78	71	8.43	96	9.80
22	4.69	47	6.86	72	8.48	97	9.85
23	4.80	48	6.93	73	8.54	98	9.90
24	4.90	49	7.00	74	8.60	99	9.95
25	5.00	50	7.07	75	8.66	100	10.00

Rule 17: The basic rules for square root are:

A. $\sqrt{\dfrac{a}{b}} = \dfrac{\sqrt{a}}{\sqrt{b}}$

B. $\sqrt{ab} = \sqrt{a}\ \sqrt{b}$

(Break down a large number into two factors, one of which can be square rooted equally.)

The steps for performing operations with roots are:

Step 1. Write the problem.

Step 2. Determine the correct rule to use and perform the mathematical problem.

Step 3. Check the answer by squaring the result.

Examples: Calculate $\sqrt{\dfrac{4}{25}}$; $\sqrt{160}$; $\sqrt{27}$.

Step 1. Write the problem:

Example 1	Example 2	Example 3
$\sqrt{\dfrac{4}{25}}$	$\sqrt{160}$	$\sqrt{27}$

Step 2. Determine the correct rule to use and perform the mathematical operation:

Rule 17A	Rule 17B	Table
$\dfrac{\sqrt{4}}{\sqrt{25}}$	$\sqrt{16} \times \sqrt{10}$	
	$4 \times \sqrt{10}$	
	4×3.16	
$\dfrac{2}{5}$	12.64	5.20

RATIO, PROPORTION, PERCENTAGE, POWERS AND ROOTS

Step 3. Check the answer:

>**Example 1:** 2/5 × 2/5 = 4/25
>**Example 2:** 12.64 × 12.64 = 159.77
>**Example 3:** 5.20 × 5.20 = 27.04

Example: If the formula for the diameter of an engine cylinder is $c = \dfrac{\sqrt{HP \times 2.5}}{N}$ **what cylinder diameter will be necessary for a 325 hp, eight-cylinder engine? In the formula, c = cylinder diameter in inches; HP = engine horsepower; and N = number of cylinders.**

Step 1. Write the problem:

$$c = \frac{\sqrt{HP \times 2.5}}{N}$$

$$HP = 325 \text{ hp}$$
$$N = 8 \text{ cylinders}$$

Step 2. Perform the mathematical operations. Note: The square root of 32.5 comes from the table, using the value halfway between 32 and 33.

$$c = \frac{\sqrt{325 \times 2.5}}{8}$$

$$c = \frac{\sqrt{25 \times 13 \times 2.5}}{8}$$

$$c = \frac{5\sqrt{13 \times 2.5}}{8}$$

$$c = \frac{5\sqrt{32.5}}{8}$$

$$c = \frac{5 \times 5.7}{8}$$

$$c = \frac{28.5}{8}$$

$$c = 3.56 \text{ inches}$$

Step 3. Check the square root calculations:

$$325 \times 2.5 = (28.5)^2$$
$$812.5 = 812.25$$

Example: Solve the problem $\sqrt{.176}$.

Step 1. Write the problem:

$$\sqrt{.176}$$

Step 2. Perform the mathematical operation:

$$\sqrt{\frac{176}{1000}}$$

$$\frac{\sqrt{176}}{\sqrt{1000}}$$

$$\frac{\sqrt{16}}{\sqrt{100}} \times \frac{\sqrt{11}}{\sqrt{10}}$$

$$\frac{4 \times \sqrt{11}}{10 \times \sqrt{10}}$$

$$\frac{4 \times 3.32}{10 \times 3.16}$$

$$\frac{13.28}{31.6} = .42$$

Step 3. Check the answer:

$$.42 \times .42 = .1764$$

Chapter 5

The Metric System and Fire Service Hydraulics

The shift from the United States from the English system of management to the International Metric System has begun. In order to prepare fire service personnel for this changeover, the impact of metrics on hydraulics will be discussed. However, the development of water movement formulas throughout this book will use the current measurement system. As the use of the metric system becomes more common, the text will be revised to reflect the change.

Although the metric system was approved by Congress in 1866 for use in the United States, obviously very little use has been made of this authorization. The National Bureau of Standards has recommended a gradual phase-in program, which is being partially implemented.

The official name for the metric units is the International System, which is abbreviated SI.

The basic units of measure for the SI system are:

length	meter
mass	gram
force	dyne or newton
liquid volume	liter
temperature	celsius
time	second
pressure	pascal
quantity of heat	joule

One of the big advantages of the SI system is that multiples of 10 of the basic unit are used. Thus, a common prefix is developed for each measurement. These prefixes are:

nano	n	1/1,000,000,000	0.000 000 001	10^{-9}
micro	μ	1/1,000,000	0.000 001	10^{-6}
milli	m	1/1000	0.001	10^{-3}
centi	c	1/100	0.01	10^{-2}
deci	d	1/10	0.1	10^{-1}
hecto	h	100	100.0	10^{2}
kilo	k	1000	1000.0	10^{3}
mega	M	1,000,000	1,000,000.0	10^{6}
giga	G	1,000,000,000	1,000,000,000.0	10^{9}

For example, if the meter is the basic unit of length, then 1000 meters would equal 1 kilometer, while one hundredth of a meter would be equal to a centimeter. The same is true of a liquid volume measure. If a pump delivers 1000 liters per minute, then this is the same as 1 kiloliter per minute.

Definitions

When compared to the British Imperial System that was established on a very arbitrary basis (an acre equals the amount of land a yoke of oxen can plow in one day), the SI sytem has logical interrelations.

meter	one ten-millionth of the distance from the equator to either pole
gram	weight of 1 cubic centimeter of water at 4°C
liter	volume of kilogram of water at 4°C
dyne	force required to produce a velocity of 1 centimeter per second when acting for 1 second on a mass of 1 gram
gram calorie	quantity of heat required to raise 1 gram of water 1 degree celsius

Abbreviations

Term	Symbol	Measurement
millimeter	mm	0.001 meter
centimeter	cm	0.01 meter
meter	m	1.0 meter
kilometer	km	1000 meters
cubic centimeters	cc	1 cm × 1 cm × 1 cm
milligram	mg	0.001 gram
centigram	cg	0.01 gram
gram	g	1.0 gram
kilogram	kg	1000 grams
milliliter	ml	0.001 liter
centiliter	cl	0.01 liter
liter	l	1.0 liter
kiloliter	kl	1000 liters
newton	N	
second	s	
pascal	Pa	N/m²
joule	J	N × m

CONVERSION

Length

1 inch	= 25.4 mm	1 mm	= 0.0394 in
	= 2.54 cm	1 cm	= 0.394 in
1 foot	= 304.80 mm		= 0.033 ft
	= 30.48 cm	1 m	= 39.37 in
	= 0.305 m		= 3.28 ft
1 mile	= 1609.34 m	1 km	= 3280.83 ft
	= 1.609 km		= .621 mi

Area

1 in²	= 645.16 mm²	1 cm²	= 0.155 in²
	= 6.45 cm²		= 0.0011 ft²
1 ft²	= 929.03 cm²	1 m²	= 10.764 ft²

Volume

1 in³	= 16.38 cc	1 cc	= 0.061 in³
	= 0.016 l	1 l	= 61.02 m³
1 ft³	= 28.32 l		= 0.035 ft³
1 gallon	= 3.79 l		= 0.264 gallons
		1 m³	= 35.31 ft³
			= 264.19 gallons

Pressure

1 psi	= 0.703 g/mm²	1 g/cm²	= 0.0142 psi
	= 70.307 g/cm²		= 2.048 lb/ft²
	= 0.0703 kg/cm²	1 kg/cm²	= 14.223 psi
	= 6.894 kPa		= 0.968 atmospheres
1 atmosphere	= 1.034 metric atmosphere	1 kPa	= .145 psi

Force

1 lb	= 444,800 dynes	1 g	= 0.0022 lb
	= 453.6 g	1 kg	= 2.205 lb
	= 4.448 N	1 N	= .225 lb

Temperature

1°F = 9/5°C + 32   1°C = 5/9(°F − 32)

Energy

1 Btu	= 1055 J	1 J	= 0.000948 Btu

Flow

1 gpm	= 0.06398 l/s	1 l/s	= 15.85 gpm

COMMON MEASUREMENTS

Pumper Ratings

750 gpm	47.31 l/s	50
1000 gpm	63.08 l/s	60
1250 gpm	78.85 l/s	80
1500 gpm	94.62 l/s	95
2000 gpm	126.16 l/s	125

Hand Lines

20 gpm	1.26 l/s	1
30 gpm	1.89 l/s	2
100 gpm	6.31 l/s	6
120 gpm	7.57 l/s	8
200 gpm	12.62 l/s	13
250 gpm	15.77 l/s	15

Pressure

Nozzle			
fog	100 psi	689.4 kPa	700 kPa
straight hand	50 psi	344.7 kPa	350 kPa
straight master stream	80 psi	551.5 kPa	550 kPa

Friction loss in 100 feet 2½-inch hose

	100 gpm	3 psi	20.68 kPa	20 kPa
	200 gpm	10 psi	68.94 kPa	70 kPa
	300 gpm	21 psi	144.77 kPa	150 kPa
	400 gpm	36 psi	248.15 kPa	250 kPa
	500 gpm	55 psi	379.17 kPa	375 kPa
head*	0.434 psi/ 1 foot		9.81 kPa/m	10 kPa/m

*Amount of pressure at base of a column of water is a gain or loss depending on position of nozzle in relation to the pump.

Conclusions

It will be some time before the fire service adopts the SI system. Standardization must be established so that new appliance sizes and hose lengths can be determined, and certain questions must be answered, such as:
1. Will the standard length of hose become 20 meters?
2. Will the unit of pressure be the kilopascal or the bar?
3. Will 60-mm diameter hose replace standard 2½-inch hose?
4. Will 75-mm diameter hose replace standard 3-inch hose?
5. Will 15 l/s be accepted as the standard hand line flow?

Once fire service personnel become familiar with the new numbers, hydraulic calculations will become much simpler than those presently in use.

Chapter 6

Fireground Hydraulics

Before jumping into the calculations and concerns of hydraulics, an understanding of the history of firefighting might be of value. A study of past firefighting techniques not only provides a proper perspective as to the importance of present equipment and techniques, but forms a foundation for further progress and technological advances.

Man, from the beginning of recorded history, has been concerned with fire. While it provided him with warmth against the cold, heat for cooking, and light for security, it was also a terrible enemy which destroyed whole cities.

In his battle to keep this force under his control, man soon learned that he had an ally with water. Early man found water ideal for fire control because there was an ample supply, it extinguished flames and it did not harm the user.

As civilization progressed, however, man found that a ready supply of water was not available when and where he needed it. Firefighting could not be accomplished in the cities where adequate water supplies did not exist. Therefore, aqueducts were developed. About 300 B.C., an aqueduct, capable of supplying over 50 million gallons of water per day, was constructed for Rome.

However, while water for domestic use now could be conveniently carried, the situation was quite different when water was needed for firefighting. In firefighting, speed was essential, and carrying heavy buckets of water from the source to the fire scene proved inefficient. From this need, a mechanical device, a pump, was developed. The pump could move the water to the fireground and enable early firefighters to deliver the water to greater heights than they could be throwing the buckets of water at the fire.

Fluids and flows

The movement of these volumes of water led to the need for understanding a new branch of science—hydraulics. The science of hydraulics is concerned with the study of fluids at rest, called hydrostatics, and fluids in motion, called hydrokinetics. A fluid is defined as a substance that yields to the slightest force and recovers its previous state when the force is removed.

Both hydrostatics and hydrokinetics are important to the firefighter because he must know what to expect when water in a hose line is not flowing (static) and what happens when an outlet is opened and movement begins (kinetic).

The origin of the science of fluids is unknown, but there are historical records that indicate that man had a working, if not scientific, knowledge of fluids as far back as 4000 B.C.

Two of the outstanding scientists in the field of hydraulics, Vitruvius and Frontinaus, issued their work about the year 97 A.D. Very little additional work appeared from this time until the engineer Leonardo da Vinci (1452-1519) published his treatise "On the Motion and Measurement of Water." He introduced a new era when he designed and had constructed the earliest chambered navigation and metering locks near Milan. This development introduced four centuries of canal building.

About 1585, Stevenius of Bruges published a paper that closely followed the work of Archimedes. The first demonstration of the pressure laws relating to fluids pressing on the base and sides of a container appeared in this paper. The principles set forth by Stevenius are:

1. Fluid pressure is perpendicular to any surface on which it acts.
2. The downward pressure of a liquid in an open vessel is proportional to its depth.

Defining pressure was a very important concept for the early hydraulic scientist, for it allowed mathematical calculations to be made. This, in turn, permitted theories to be changed to practical applications.

In 1730 Henri Pitot invented the tube which bears his name and is used today, in a refined form, for measuring velocities in pipes and open channels.

A method for computing the quantity of flow was developed by G. B. Venturi around 1798. His studies on the behavior of fluids when passing through a constriction formed the basis for the development of the venturi meter. The modern venturi meter is used in municipal water works stations, industrial sites, and as a research tool.

In 1888, John R. Freeman conducted a series of hydraulic studies that had far-reaching effects on the fire service. He studied the flow of water in ordinary fire hose of various roughnesses and qualities and experimented with nozzles under varying pressures and diameters. Freeman noted the characteristics of the water stream and the heights and distances to which they were carried, both vertically and horizontally. In addition, he developed a standard nozzle (the Underwriters playpipe) that could be used as an accurate water meter.

Today, no matter how many pieces of apparatus are on the fireground, and no matter how many firefighters are available, the problem is still the same—delivering the water from the source to the fire. Almost any type of carrier, from a bucket to a hose line, can be used. As far as the firefighter is concerned, the most frequently used method is the hose line. As soon as movement of water through a restrained carrier is attempted, a knowledge of the principles of hydraulics is required.

Maximum efficient of pumpers and appliances is dependent upon the principles discussed in this chapter.

"The fire apparatus driver/operator shall demonstrate the principles of friction loss as they relate to:

"(a) Internal diameter of hose
"(b) Length of hose line
"(c) Manner in which hose lines are laid
"(d) Physical condition of hose
"(e) Pressure
"(f) Use of appliances
"(g) Use of multiple hose lines
"(h) Use of various nozzles
"(i) Velocity of flow.

"The fire apparatus driver/operator shall identify the following types of fluid pressure encountered in the fire service:

"(a) Flow pressure
"(b) Negative pressure
"(c) Normal operating pressure
"(d) Residual pressure
"(e) Static pressure.

"The fire apparatus driver/operator shall identify the following terms that relate to the basic principles of fire service hydraulics:

"(a) Atmospheric pressure
"(b) Capacity
"(c) Displacement
"(d) Flow (gpm)
"(e) Friction loss
"(f) Head pressure (back pressure)
"(g) Hydrant pressure
"(h) Net engine pressure
"(i) Nozzle reaction
"(j) Pounds per square inch
"(k) Pump discharge pressure
"(l) Vacuum
"(m) Velocity
"(n) Water hammer.

"The fire apparatus driver/operator, given a series of fireground situations and using the written formulas specified by the authority having jurisdiction, shall determine:

"(a) Nozzle or pump discharge pressures when the length and size of hose, and size of nozzle are given.

"(b) Water flow in gallons per minute when the diameter of the orifice and pressure at the orifice are given.

"(c) The friction loss in the supply and attack lines, used by the authority having jurisdiction, when the gpm flow is given.

"(d) Friction loss in siamesed lines when size of hose and gpm flow are given.

"(e) Friction loss in wyed lines when size of hose and gpm flow are given.

"(f) Friction loss in multiple lines when the size of hose and gpm flow are given.

"(g) An estimated remaining available volume from a hydrant while pumping a given volume.

"The fire apparatus driver/operator, given a series of fireground situations, shall calculate correct pump discharge pressure, gpm, friction loss, and nozzle pressure, using mental formulas specified by the authority having jurisdiction.

"The fire apparatus driver/operator, given a fire department pumper and a series of fireground situations, shall produce effective hand and master streams specified by the authority having jurisdiction.

"The fire apparatus driver/operator, given a selection of nozzles and tips, shall identify the type, design, operation, nozzle pressure, and flow in gpm for proper operation of each."*

PRESSURE

The compressive forces used in the fire service, when referenced to a given area, are known as *pressure*. Pressure denotes a force per unit area, with the measurement of force in pounds and the measurement of area in either square inches or square feet.

It is extremely important to distinguish between a force and a pressure.

A *force* is a weight, and whether this weight acts on a small area or a large area determines the amount of pressure.

---

*Paragraphs 3-4.1, 3-4.2, 3-4.3, 3-4.8, 3-4.9, 3-6.2, and 3-6.10. Reprinted with permission from NFPA 1002-1982, Standard for Fire Apparatus Driver/Operator Professional Qualifications, Copyright© 1982, National Fire Protection Association, Quincy, Massachusetts 02269. This reprinted material is not the complete and official position of the NFPA on the referenced subject, which is represented only by the standard in its entirety.

If a 200-pound weight were placed upright on a 4-square-inch platform (figure 1), the pressure exerted would be:

$$\text{pressure} = \frac{\text{force}}{\text{unit area}} = \frac{200 \text{ pounds}}{4 \text{ square inches}} = 50 \text{ psi}$$

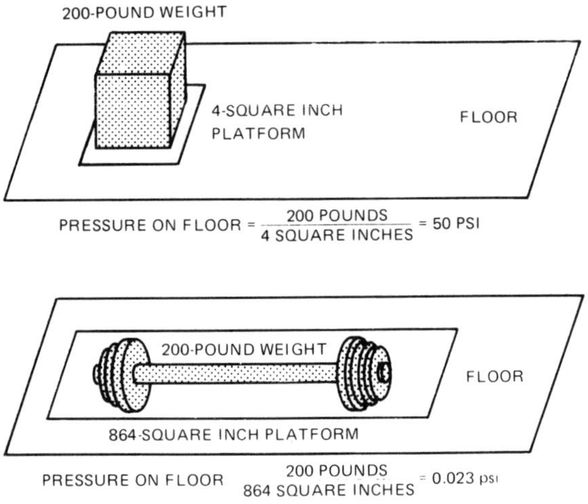

Figure 1. Force versus pressure.

Now, if this same weight were to be placed horizontally on an 864-square-inch platform, the pressure exerted would be:

$$\text{pressure} = \frac{\text{force}}{\text{unit area}} = \frac{200 \text{ pounds}}{864 \text{ square inches}} = .023 \text{ psi}$$

The same force, 200 pounds, is present, but by redistributing it over a different size area, the pressure changes. Sometimes, in a sort of slang expression, pressure is referred to only as pounds. Remember, when this happens, the square inch area is understood, even if it is not mentioned.

Types of pressure

Within the fire service, there are several types of pressure that are used consistently:

Atmospheric pressure is the pressure caused by the weight of the air. This pressure will vary depending upon the elevation at which it is being measured. If a 1-square-inch section of the atmosphere were cut out from sea level to outer space where it ends and then weighed, the air would weigh 14.7 pounds (figure 2A). In areas below sea level, there is a higher column of air and so the atmospheric pressure would be greater (figure 2B). On a mountain top, atmospheric pressure would be less (figure 2C). The pump operator must be familiar with the characteristics of atmospheric pressure for drafting operations, which are explained in Chapter 15.

Table 1 shows atmospheric pressures for various altitudes. However, a good rule of thumb is to use a ½-pound-per-square-inch decrease in pressure for each 1000-foot increase in altitude.

Gage pressure is the pressure read on the gage. However, atmospheric pressure is always around, yet gages on fire apparatus read 0 psi. Gages, therefore, have been adjusted to read disregarding atmospheric pressure. To indicate that a reading does not account for atmospheric pressure, the abbreviation psig (pounds per square inch gage) is used.

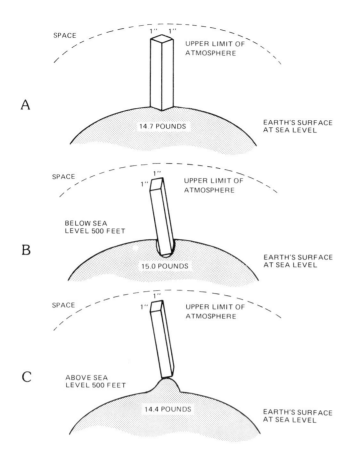

Figure 2. Atmospheric pressure.

TABLE 1. Atmospheric Pressure Versus Altitude					
Altitude feet	Pressure psi	Altitude feet	Pressure psi	Altitude feet	Pressure psi
−1000	15.2	2500	13.4	6000	11.8
−500	15.0	3000	13.2	6500	11.5
0	14.7	3500	12.9	7000	11.3
+500	14.4	4000	12.7	7500	11.1
+1000	14.2	4500	12.4	8000	10.9
1500	13.9	5000	12.2	8500	10.7
2000	13.7	5500	12.0	9000	10.5

Absolute pressure is the sum of gage pressure and atmospheric pressure. The abbreviation for absolute pressure is psia. Mathematically, the definition can be stated:

$$\text{psia} = \text{psig} + \text{atmospheric pressure}$$

At a 500-foot elevation, the absolute pressure when the gage reads 118 psig is:

$$\text{Absolute pressure} = \text{gage pressure (118 psig)} + \text{atmospheric pressure (14.4 psi)}$$

$$\text{Absolute pressure} = 118 + 14.4 = 132.4 \text{ psia}$$

Back pressure or head pressure is the pressure created by a column of water due to elevation. As in atmospheric pressure, back pressure is created by the weight of the water above the measurement point. Methods for calculating back pressure are explained in this chapter.

Positive pressure is pressure above atmospheric pressure. Since a 0 psig reading on the gage indicates atmospheric pressure, any reading above 0 on the gage would be a positive pressure.

Negative pressure is pressure above atmospheric. This pressure then, is a partial vacuum (absence of air) until at 0 psia there would be a perfect vacuum. The relationship between positive and negative pressures is shown in figure 3.

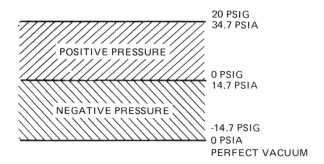

Figure 3. Positive and negative pressures.

Static pressure is the pressure when the water is not moving. It is energy that is available but not being used. For example, suppose a pumper connects to a hydrant, opens the hydrant, and supplies water to the pump. The intake gage on the pumper reads 60 psig. With no water being discharged from the pump, the 60-psig reading is a static pressure reading.

Residual pressure is the pressure remaining once water has begun flowing. Using the preceding example, the static pressure reading is 60 psig. Now if water is discharged, the intake reading will drop. This new reading, which might be 54 psig, is called residual pressure. The difference between the static reading and the residual reading is due to losses required to overcome friction and elevation.

Flow pressure is the pressure available to move water from one point to another.

Normal operating pressure is the pressure which has been established by a fire department for specific, commonly encountered fireground situations. The normal operating pressure for a 1½-inch, 200-foot hand line with a fog nozzle is 160 psig.

Pump discharge pressure is the pressure read on the master discharge gage, read in psi, and indicating the amount of pressure being created by the pump.

Hydrant pressure is the pressure available from a hydrant under flowing conditions. It is also used to denote a hand line that is connected directly to the hydrant and operating only from the pressure availble from the hydrant.

Net engine pressure is the pressure that is actually produced by the pump. It is the difference between the intake pressure and the discharge pressure. For example, if a hydrant is supplying 60 psi pressure and there is a discharge pressure of 150 psi, the pump is actually producing only 90 psi, which is the net engine pressure.

FRICTION LOSS

Movement of fluids through a conduit causes a loss of energy. The amount of energy applied by the pump and recorded as engine pressure will not be the amount of energy reaching the nozzle. There will be energy losses due to many factors, but these losses are generally combined together and called friction loss.

The factors that influence friction loss are:
1. The type of flow — laminar or turbulent.
2. The quantity of water being pumped.
3. The diameter of the hose.
4. The length of the hose.
5. The quality and age of the hose.
6. Appliances attached to the hose stretch, i.e., siamese and wye.

Laminar and turbulent flow

As water flows through a pipe or a hose line, it can be either turbulent or laminar. In *laminar flow*, the water moves along in straight lines. This movement would appear to be layered, with one layer of water stacked on top of another, all moving along together (figure 4A). Only a small amount of water would be in contact with the wall of the conduit.

Since the layers of water at the edge of the conduit (water layers 1 and 7) are touching the wall, they will move slower than the layers just next to them (water layers 2 and 6). This will cause layers 2 and 6 to slow down. Now, layers 3 and 5 will move just a little faster, since they are only in contact with other water layers. The center layer, 4, will move the fastest. (It is important to remember that these layers are extremely thin and cannot actually be seen.) A graph of the velocity at a point in the hose line is shown in figure 4B. Layer 4 is shown as having the greatest velocity, least friction, by using the longest arrow.

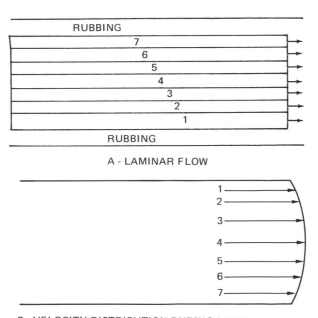

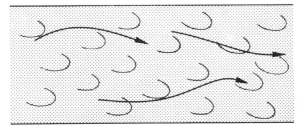

Figure 4. Water flow in a conduit.

As pressure increases, the water stops moving in nice, even layers and begins to swirl around, moving in one general direction. The swirling action (figure 4C) starts at the surface of the conduit and then affects the interior layers. This type of flow is called *turbulent flow*.

Quantity of water flowing

With a given size hose, the greater the quantity of water pumped, the higher the friction loss. Flow is measured in hundreds of gallons per minute and is usually represented with the letter Q (Q = gpm ÷ 100). Mathematically, friction loss will increase approximately as the square of the discharge (Q^2). Again, this is an approximate figure because the formula is not exact. In addition to the quality of the hose, water flow also depends on the type and size of the nozzle, and the friction loss will vary as nozzle, hose and pressure factors change.

Example: If the flow through a hose line went from 200 to 600 gpm, how much would the friction loss increase?

Since the flow has tripled, friction loss will increase Q^2 or $3^2 = 9$ times over what it was when only 200 gpm were flowing.

Diameter of hose

The larger the diameter of the hose, the less the friction loss will be with the same quantity of water flowing. Mathematically, it can be stated that friction loss varies inversely as the fifth power of the diameter.

$$\text{friction loss} = \frac{1}{d^5}$$

Example: If the diameter of a hose is tripled, how much would the friction loss decrease with the same quantity of water flowing?

The friction loss will decrease as the fifth power:

$$\text{friction loss} = \frac{1}{3 \times 3 \times 3 \times 3 \times 3} = \frac{1}{243}$$

So, if the original friction loss had been 243 psi and the hose diameter was tripled, the new line would have a friction loss of

$$\frac{1}{243} \times 243 = 1 \text{ psi}$$

for the same quantity of water flowing.

Another way of reducing the friction loss is to increase the number of lines. This, in effect, increases the diameter of hose available to carry the water.

Length

The length of the hose line directly influences the amount of friction loss. The longer the hose line, the greater the friction loss. If the length is increased three times, the friction loss also triples, if the discharge remains constant.

Example: If the friction loss in 400 feet of hose is 60 psi, what will be the friction loss in 500 feet of the same size hose flowing the same amount of water?

A proportion can be formed:

$$\frac{400}{60} = \frac{500}{X}$$

$$400X = 500 \times 60$$

$$X = \frac{500\ (60)}{400}$$

$$X = \frac{300}{4}$$

$$X = 75 \text{ psi}$$

Also with this area, friction loss is affected by the manner in which the hose lines are laid. The more bends, the greater the friction loss. The fireground calculations take this fact into consideration, so it is only necessary to use the length of the lay.

Quality and age of hose

New polyester hose jackets with rubber lining will have a different friction loss than jackets made of cotton and lined with rubber. Even if the material for the jackets is the same, the tightness of the weave will affect friction loss.

As pressure increases in a hose line, the rubber lining is forced into the woven jacket (figure 5A). The rubber liner then assumes the shape of the jacket weave, the looser the jacket weave, the rougher the inner liner under pressure. This increased roughness causes an increase in friction loss. Increased pressure also tends to elongate the hose (figure 5B). The elongation provides a different shaped surface, thereby increasing friction loss.

As hose ages, the rubber liner tends to get rougher. In addition, as the hose is used, foreign objects in the water supply system cause deterioration of the liner.

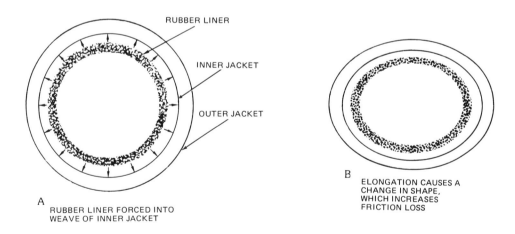

Figure 5. Water hose under pressure.

Appliances

Appliances such as siameses, wyes, and ladder pipes all add to the total friction loss because they restrict the waterway and disturb the smooth flow. The amounts of friction losses added by these devices are discussed later in this chapter.

Additives

Additives such as Rapid Water, a development of the Union Carbide Corporation, reduce the friction loss in hose. These chemical additives tend to change the flow characteristics in such a way that friction loss is reduced. Tests have shown that 1 gallon of Rapid Water injected into 6000 gallons of water will increase water delivery by 35 to 45 percent. The optimum injection is 1 gallon added to 3000 gallons of water, resulting in an increased delivery of 50 to 65 percent. Naturally, calculations and formulas assume no additives in the water.

HYDRAULIC CALCULATIONS

The entire objective of hydraulic calculations is to permit the pump operator to determine the necessary engine discharge pressure for operating at the scene. The basic equation for this is:

$$EP = NP + FL \pm E, \text{ where}$$
EP = engine pressure in psi
NP = nozzle pressure in psi
FL = friction loss in psi
$+E$ = elevation in psi necessary to overcome pressure exerted by height of discharge above the pump
$-E$ = reduction in psi, where the discharge is below the pump

So, the pump operator only needs to calculate each of these values. The techniques for accomplishing these calculations are explained in the following sections.

Nozzle pressure

Depending on the type of nozzle and its intended use, there are three main nozzle pressures that must be considered. All fog nozzles, 1½, 2½ and master stream, can be considered to operate at maximum efficiency at 100 psi nozzle pressure (figure 6). Smooth bore nozzles used as hand lines should be operated at 50 psi, while smooth bore nozzles used for master streams are designed to operate at 80 psi (figure 7).

A. 1½-inch, 100-gpm, 100-psi Elkhart Brass.

B. 2½-inch, 240-gpm, 100-psi Elkhart Brass.

C. Master stream, 1000-gpm, 100-psi Elkhart Brass.

Figure 6. Fog nozzles.

Friction loss calculations for 2½-inch lines

Now, with the knowledge on how to calculate flow, values of friction loss can be determined. There are two basic formulas used for calculating friction loss in 2½-inch hose lines, the basic size used in hydraulic calculations.

A. Elkhart Brass smooth-bore hand nozzle, 50 psi nozzle pressure. Nozzle diameter is changed by changing tip.

B. Elkhart Brass smooth-bore master stream nozzle, 80 psi nozzle pressure. Nozzle diameter is changed by changing tip.

Figure 7. Smooth-bore nozzles.

$FL = 2Q^2 + Q$, where
FL = friction loss in 100 feet of 2½-inch hose
Q = flow in gallons per minute divided by 100

Presently, there is some disagreement within the fire service concerning the accuracy of this formula. Some proponents claim that too high a figure is obtained if this formula is used. However, in tests recently conducted, it has been my experience that this formula still approximates the friction loss in 2½-inch hose. New hose did show somewhat less loss, but the same brands that were a few years old again had a friction loss that approximated this formula. In addition, some allowance must be made for the manner in which the hose lines are laid. Friction loss calculations in this book will use this equation.

Equation: What is the friction loss in 400 feet of 2½-inch hose flowing 300 gpm?

Step 1. Select the proper equation when flow is known and above 100 gpm:

$$FL = 2Q^2 + Q$$

Step 2. Determine the formula values:

$$Q = \frac{300}{100} = 3$$

Step 3. Solve the equation:

$$FL = 2(3)^2 + 3$$
$$FL = 2(9) + 3$$
$$FL = 21 \text{ psi}/100 \text{ feet}$$
$$FL = 21 \times 4 = 84 \text{ psi}/400 \text{ feet}$$

Friction loss calculations for small flows

The formula for calculating friction loss for flows that are less than 100 gpm is:

$FL = 2Q^2 + \frac{1}{2}Q$, where
FL = friction loss in 2½-inch hose per 100 feet
Q = flow in gpm divided by 100

PUMP OPERATORS HANDBOOK

Example: What is the friction loss in 300 feet of 2½-inch hose flowing 80 gpm?

Step 1. Select the proper equation for a flow less than 100 feet:

$$FL = 2Q^2 + \tfrac{1}{2}Q$$

Step 2. Determine the formula values:

$$Q = \frac{80}{100} = .8$$

Step 3. Solve the equation:

$$FL = 2(.8)^2 + \tfrac{1}{2}(.8)$$
$$FL = 2(.64) + .4$$
$$FL = 1.68 \text{ psi}/100 \text{ feet}$$
$$FL = 1.68 \times 3 = 5.04 \text{ psi}/300 \text{ feet}$$

Friction loss for 1½, 3, 3½ and 4-inch hose

One of the easiest ways to calculate the friction loss in hose other than 2½-inch is the following:

Step 1. Calculate the friction loss as if the hose were 2½-inch.

TABLE 2. Friction Loss in Fire Hose

GPM	3/4-inch	1-inch	1½-inch	1¾-inch	2-inch	2½-inch	2½-inch Coupling	3-inch 3-inch Coupling
10	13.5	3.5	.3					
15	29.0	7.2	.7					
20	50.0	12.3	1.7					
25	75.0	18.5	2.6					
30	105.0	26.0	3.7					
35	142.0	35.0	4.8					
40		44.0	6.2					
45		55.0	7.6					
50		67.0	9.4					
60		94.3	13.1	4.3	2.2			
70		126.0	17.5	5.9	3.1			
80			22.5	7.6	3.9			
90			27.8	9.0	4.6			
100			33.5	11.1	5.7	2.5	1.2	1.2
110			37.6	13.4	6.8	3.0	1.4	1.4
120			45.5	15.3	7.8	3.6	1.6	1.6
130			54.5	20.7	10.2	4.3	1.9	1.8
140				26.3	13.1	5.2	2.1	2.0
150				31.5	16.2	5.8	2.5	2.3
160				36.4	18.5	6.6	2.9	2.6
170				41.2	20.8	7.4	3.2	2.9
180				45.2	22.9	8.3	3.6	3.2
190				49.0	25.0	9.2	3.8	3.5
200				53.0	27.2	10.1	4.0	3.8
210					29.6	11.1	4.4	4.2
220					32.0	12.0	4.6	4.6
230					35.2	13.0	5.3	5.0
240					38.2	14.1	5.8	5.4
250					40.8	15.3	6.2	5.9
260						16.4	6.8	6.3
270						17.5	7.3	6.7
280						18.7	7.8	7.2
290						19.9	8.4	7.7

TABLE 2. Friction Loss in Fire Hose *continued*						
GPM	2½-inch	3-inch 2½-inch Coupling	3-inch 3-inch Coupling	Siamese 2½ & 3	3½-inch	4-inch	5-inch
300	21.2	9.0	8.2		3.7	2.2	
310	22.5	9.7	8.7		4.0	2.5	
320	23.8	10.3	9.3		4.2	2.6	
330	25.3	10.9	9.9		4.4	2.7	
340	26.9	11.6	10.5		4.7	2.8	
350	28.4	12.3	11.0		5.0	2.9	
360	30.0	13.0	11.5		5.2	3.2	
370	31.5	13.7	12.2		5.5	3.2	
380	33.0	14.4	12.8		5.8	3.3	
390	34.6	15.2	13.4		6.0	3.5	
400	36.2	16.0	14.1	5.9	6.3	3.7	
420	40.0	17.7	15.4	6.5	6.8	4.0	
440	43.8	19.4	16.8	7.1	7.4	4.4	
460	47.2	21.3	18.2	7.7	8.0	4.7	
480	51.1	23.1	19.7	8.3	8.7	5.1	
500	55.0	25.0	21.2	9.0	9.5	5.5	2.0
525			23.2		10.5		
550			25.2		11.4		
575			27.5		12.4		
600			29.9	12.7	13.4	7.7	2.6
625			32.0		14.4		
650			34.5		15.5		
675			37.0		16.6		
700			39.5	16.8	17.7	10.3	3.5
725			42.3		18.9		
750			45.0		20.1		
775			47.8		21.4		
800			50.5	21.5	22.7	13.2	4.4
825			53.5		24.0		
850			56.5		25.4		
875			59.7		26.8		
900			63.0	26.7	28.2	16.4	5.8
950					31.2		
1000					34.3	19.9	6.7
1050					37.8		
1100					41.0	23.8	8.0
1150					44.6		

Step 2. Multiply the 2½-inch friction loss by the following factors:

1½-inch — 11
3-inch with 3-inch couplings — .38
3-inch with 2½-inch couplings — .4
3½-inch — .17
4-inch — .091

Step 3. Multiply the friction loss of step 2 by the length of the hose line in hundreds of feet.

Example: What is the friction loss when 300 gpm is flowing in 300 feet of 1½-inch hose and 4-inch hose?

Step 1. Calculate the friction loss for 2½-inch line:

$$FL = 2Q^2 + Q$$
$$FL = 2(3)^2 + 3$$
$$FL = 21 \text{ psi}$$

Step 2. Multiply by the conversion factors: 1½-inch — 11; 4-inch — .091:

$$FL_{1½} = 21 \times 11$$
$$= 231 \text{ psi}/100 \text{ feet}$$
$$FL_4 = 21 \times .091$$
$$= 1.9 \text{ psi}/100 \text{ feet}$$

Step 3. Multiply by the length:

$$FL_{1\frac{1}{2}} = 231 \times \frac{300}{100}$$
$$= 693 \text{ psi}/300 \text{ feet}$$
$$FL_4 = 1.9 \times \frac{300}{100}$$
$$= 6 \text{ psi}/300 \text{ feet}$$

As with many other calculations in hydraulics, friction loss can be determined in advance. Table 2 contains the common friction losses encountered in the fire service. These values are more accurate than those calculated on the fireground because of the inaccuracies introduced into the formulas in order to simplify them.

Friction loss in unequal lengths of hose

Sometimes when dual lines are laid, they will be of unequal lengths. When this happens, friction loss calculations can be based on the average length of the hose lay. For example, if one line is 200 feet long and the other is 300 feet, use the average length in calculating friction loss.

$$\frac{200 + 300}{2} = \frac{500}{2} = 250 \text{ feet}$$

So, if friction loss per hundred feet is 21 psi, the loss in 250 feet is

$$21 \times \frac{250}{100} = 21 \times 2.5 = 52.5 \text{ psi}/250 \text{ feet}$$

Friction loss in devices

Water flowing through large appliances such as ladder pipes and deck guns will produce additional friction losses. These friction losses will vary with the type of device, amount of water flowing, manufacturer, and age.

Each fire department should run flow tests to determine the friction loss in its large-gallonage devices. For purposes of problem solving use 10 psi loss for ladder pipes, deck guns and wagon pipes; use 5 psi loss for each siamese and wye inserted in the line.

For elevating platform streams, the friction loss is dependent upon the pipe size from the manifold to the basket. For friction loss in this pipe size refer to table 3.

Friction loss in parallel lines

Friction loss in parallel lines can be determined by dividing the water equally in the hose lines, provided the hose lines are of the same diameter, and then calculating the friction loss for a single line. For example, if the flow through two 2½-inch lines is 600 gpm, then the flow through each 2½-inch line is 300 gpm. The friction loss can then be calculated based upon 300 gpm flowing through a single 2½-inch line.

Another method for calculating friction loss in parallel lines is to calculate the friction loss as if the entire amount of water was flowing through a single 2½-inch line and then multiply the answer by the following factors:

two 2½-inch lines — 1/4
three 2½-inch lines — 1/8
one 2½-inch and one 3-inch line — 1/6
two 3-inch lines — 1/9

FIREGROUND HYDRAULICS

TABLE 3. Friction Loss in Pipe

Pipe Size	Flow	Friction Loss per 100 Feet
4-inch	500	5.6
	600	8.7
	700	10.8
	800	14.1
	900	17.7
	1000	21.8
5-inch	500	1.8
	600	2.6
	700	3.4
	800	4.4
	900	5.6
	1000	6.8
6-inch	500	0.7
	600	1.0
	700	1.4
	800	1.7
	900	2.2
	1000	2.7

Example: What is the friction loss in each of three parallel 2½-inch hose lines of 700 feet each, if the total flow is 800 gpm?

Step 1. Calculate the friction loss for 2½-inch hose:

$$FL = 2Q^2 + Q$$
$$FL = 2(8)^2 + 8$$
$$FL = 2(64) + 8$$
$$FL = 136 \text{ psi}/100 \text{ feet}$$

Step 2. Multiply by the conversion factor 1/8:

$$FL_{3-2\frac{1}{2}} = 136 \times 1/8$$
$$= 17 \text{ psi}/100 \text{ feet}$$

Step 3. Multiply by the length:

$$FL_{3-2\frac{1}{2}} = 17 \times \frac{700}{100}$$
$$= 119 \text{ psi}/700 \text{ feet}$$

Friction loss simplification for the fireground

For ease of operation, a simple method of calculating friction loss in 100 feet of 2½-inch hose has been developed. This method, known as the hand method, is shown in figure 8. Starting with the thumb, each finger is given an odd number starting with 3 (3, 5, 7, 9 and 11). Again, starting with the thumb, give each finger a discharge in hundreds of gallons per minute. For example, 1 for 100 gpm, 2 for 200 gpm, etc.

Now, using this method, for a flow of 400 gpm, the friction loss in 100 feet of 2½-inch hose would be $9 \times 4 = 36$ psi. Using this method, friction loss for 2½-inch hose is:

Flow	FL/100 Feet
100 gpm	3 psi
200 gpm	10 psi
300 gpm	21 psi
400 gpm	36 psi
500 gpm	55 psi

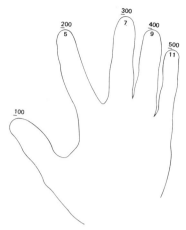

Figure 8. Hand method for computing friction loss in 2½-inch hose.

For 3-inch lines with flows up to 500 gpm, the operator can use:

$$FL_3 = Q^2$$

This yields:

Flow	FL/100 Feet
100 gpm	1 psi
200 gpm	4 psi
300 gpm	9 psi
400 gpm	16 psi
500 gpm	25 psi

Elevation

One of the laws of physics states that the pressures exerted by water at any point in an open vessel is dependent upon its depth. This pressure can be calculated by using several formulas and the physical fact that 1 cubic foot of fresh water weighs 62.5 pounds.

If 1 cubic foot of water were divided into square inches (figure 9), there would be 144 columns of water (12 by 12), each 1 foot high. Each column would also have a base area of 1 square inch (1 inch by 1 inch). The pressure at the base is solely dependent upon the weight of the water.

Since 1 cubic foot of fresh water weighs 62.5 pounds, then one of the columns, 1 square inch at the base and 1 foot high weighs

$$\frac{62.5 \text{ pounds}}{144 \text{ square inches}} = .434 \text{ pounds per square inch}$$

The significance of this value can best be shown in the following examples:

Example: What is the pressure at the bottom of a 100-foot-deep fresh water lake, allowing that water is not compressible?

For each foot of depth, water exerts a pressure of .434 pounds. At 100 feet, the pressures would be 100 times as great, so that

$$\text{Pressure} = 100 \text{ feet} \times \frac{.434 \text{ psi}}{1 \text{ ft}}$$

$$\text{Pressure} = 43.4 \text{ psi}$$

Example: The water level of a reservoir is 200 feet above a fire hydrant. What is the static reading at the hydrant?

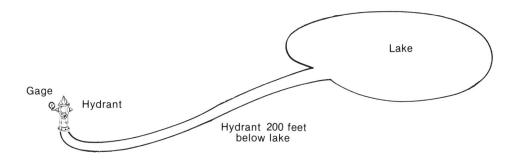

The pressure exerted by water that is elevated is again equal to .434 psi for every foot of elevation. For 200 feet of elevation the pressure would be:

$$\text{Pressure} = 200 \text{ feet} \times \frac{.434 \text{ psi}}{1 \text{ ft}}$$

$$\text{Pressure} = 86.8 \text{ psi}$$

Note that in neither of these examples was the amount of water nor the area of the water mentioned. This is because the pressure is only dependent upon the height, and with everything being referenced to 1 square inch, the area is not needed.

For fireground operation, the value .434 psi per foot of elevation can be rounded off to .5. This means that ½ psi per foot can be used. For a ladder pipe elevated 60 feet in the air, back pressure can be estimated to be ½ × 60 = 30 psi.

This rule of thumb can also be used for standpipe calculation. A value of 5 psi per story above ground level can be used for back pressure in a standpipe system. A connection four stories above ground level would have a back pressure of 4 × 5 = 20 psi.

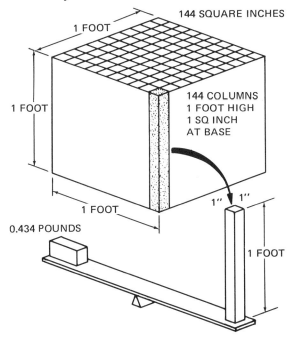

Figure 9. Pressure-height relationship.

Quantity of water flowing

On the fireground, the amount of water flowing is arrived at by means of educated guesses. With certain types of nozzles, the flow can be determined by calculation or from table 4. The flow is based upon the formula

$Q = 29.7 \times d^2 \times \sqrt{P} \times c$, where
Q = flow in gpm
d = diameter of a smooth bore nozzle
P = nozzle pressure in psi
c = coefficient of discharge that varies with the device (hydrant, nozzle)

TABLE 4. Discharge of Smooth Nozzles, Gpm

Nozzle Pressure	Nozzle Size, Inches									
	1/8	1/4	3/8	1/2	5/8	3/4	7/8	1	1-1/8	1-1/4
20	2.1	8.3	18	33	52	74	101	132	167	206
22	2.2	8.7	19	35	54	78	106	139	175	216
24	2.3	9.0	20	36	56	81	111	145	183	226
26	2.4	9.3	21	38	59	85	115	151	191	235
28	2.5	9.7	22	39	61	88	120	157	198	244
30	2.5	10.0	22	40	63	91	124	162	205	253
32	2.6	10.5	23	42	65	94	128	167	212	261
34	2.7	10.7	24	43	67	97	132	172	218	269
36	2.7	11.0	25	44	69	100	136	177	224	277
38	2.8	11.2	25	45	71	102	140	182	231	285
40	2.9	11.7	26	47	73	105	143	187	237	292
42	3.0	12.0	27	48	74	108	147	192	243	299
44	3.0	12.2	27	49	76	110	150	196	248	306
46	3.1	12.5	28	50	78	113	154	200	254	313
48	3.2	12.7	29	51	80	115	157	205	259	320
50	3.2	13.0	29	52	81	118	160	209	265	326
52	3.3	13.2	30	53	83	120	164	213	270	333
54	3.4	13.5	30	54	84	122	167	217	275	339
56	3.4	13.7	31	55	86	124	170	221	280	345
58	3.5	14.0	31	56	87	127	173	225	285	351
60	3.5	14.2	32	57	89	129	176	229	290	357
62	3.6	14.5	32	58	90	131	179	233	295	363
64	3.7	14.7	33	59	92	133	181	237	299	369
66	3.7	15.0	34	60	93	135	184	240	304	375
68	3.8	15.2	34	61	95	137	187	244	308	381
70	3.8	15.2	35	61	96	139	190	247	313	386
72	3.8	15.5	35	62	97	141	192	251	318	391
74	3.9	15.7	36	63	99	143	195	254	322	397
76	4.0	16.0	36	64	100	145	198	258	326	402
78	4.0	16.2	37	65	101	147	200	261	330	407
80	4.1	16.5	37	66	103	149	203	264	335	413
82	4.1	16.7	38	67	104	151	205	268	339	418
84	4.1	16.7	38	67	105	152	208	271	343	423
86	4.2	17.0	38	68	107	154	210	274	347	428
88	4.3	17.2	39	69	108	156	213	277	351	433
90	4.3	17.5	39	70	109	158	215	280	355	438
92	4.3	17.5	40	70	110	160	218	283	359	443
94	4.4	17.7	40	71	111	161	220	286	363	447
96	4.5	18.0	41	72	113	163	222	289	367	452
98	4.5	18.2	41	73	114	165	225	292	370	456
100	4.6	18.5	41	74	115	166	227	295	374	461
105	4.7	19.0	42	76	118	170	233	303	383	473
110	4.8	19.2	43	77	121	174	238	310	392	484
115	4.9	19.7	44	79	123	178	243	317	401	495
120	5.0	20.2	45	81	126	182	249	324	410	505
125	5.1	20.5	46	82	129	186	254	331	418	516
130	5.2	21.0	47	84	131	190	259	337	427	526
135	5.3	21.5	48	86	134	193	264	343	435	536
140	5.4	21.7	49	87	136	197	269	350	443	546
145	5.5	22.2	50	89	139	200	273	356	450	556
150	5.6	22.5	50	90	142	202	278	363	458	565

FIREGROUND HYDRAULICS

TABLE 4. Discharge of Smooth Nozzles, Gpm *(continued)*

Nozzle Pressure	1-3/8	1-1/2	1-5/8	1-3/4	1-7/8	2	2-1/4	2-1/2	3
20	250	298	350	407	468	532	674	832	1198
22	263	313	367	427	490	557	707	874	1258
24	275	327	384	446	512	582	739	913	1314
26	286	340	400	464	533	606	769	947	1368
28	297	353	415	481	554	629	799	987	1420
30	307	365	429	498	572	651	826	1021	1468
32	317	377	443	514	591	673	854	1055	1518
34	327	389	457	530	610	693	880	1088	1565
36	336	400	470	546	627	713	905	1118	1610
38	345	411	483	561	645	733	930	1150	1652
40	354	422	496	575	661	752	954	1180	1700
42	363	432	508	589	678	770	978	1213	1740
44	372	442	520	603	694	788	1000	1236	1783
46	380	452	531	617	710	806	1021	1261	1819
48	388	462	543	630	725	824	1043	1290	1856
50	396	472	554	643	740	841	1065	1316	1892
52	404	481	565	656	754	857	1087	1342	1930
54	412	490	576	668	769	873	1108	1370	1970
56	419	499	586	680	782	889	1129	1397	2004
58	426	508	596	692	796	905	1149	1421	2040
60	434	517	607	704	810	920	1168	1445	2075
62	441	525	617	716	823	936	1187	1467	2110
64	448	533	627	727	836	951	1206	1489	2145
66	455	542	636	738	850	965	1224	1512	2179
68	462	550	646	750	862	980	1242	1533	2210
70	469	558	655	761	875	994	1260	1555	2240
72	475	566	665	771	887	1008	1278	1577	2272
74	482	574	674	782	900	1023	1296	1599	2302
76	488	582	683	792	911	1036	1313	1621	2336
78	494	589	692	803	924	1050	1330	1643	2365
80	500	596	700	813	935	1063	1347	1665	2395
82	507	604	709	823	946	1076	1364	1685	2425
84	513	611	718	833	959	1089	1380	1705	2455
86	519	618	726	843	970	1102	1396	1725	2480
88	525	626	735	853	981	1115	1412	1745	2510
90	531	633	743	862	992	1128	1429	1765	2540
92	537	640	751	872	1002	1140	1445	1785	2570
94	543	647	759	881	1012	1152	1460	1805	2598
96	549	654	767	890	1022	1164	1476	1825	2625
98	554	660	775	900	1032	1176	1491	1845	2654
100	560	667	783	909	1043	1189	1506	1860	2679
105	574	683	803	932	1070	1218	1542	1908	2745
110	588	699	822	954	1095	1247	1579	1948	2805
115	600	715	840	975	1120	1275	1615	1995	2070
120	613	730	858	996	1144	1303	1649	2035	2929
125	626	745	876	1016	1168	1329	1683	2080	2995
130	638	760	893	1036	1191	1356	1717	2120	3051
135	650	775	910	1056	1213	1382	1750	2160	3114
140	662	789	927	1076	1235	1407	1780	2198	3168
145	674	803	944	1095	1257	1432	1812	2240	3225
150	686	817	960	1114	1279	1456	1843	2278	3280

Reprinted with permission from The Waterous Company, St. Paul, Minn.

The 2½-inch discharge butt of a fire hydrant will have a c value of from .7 to .9, depending upon the construction of the hydrant (figure 10). Smooth bore nozzles will have a c value ranging from .96 to .99. Since the values .96 and .99 are about 1.0, the coefficient is sometimes deleted from the calculations for smooth bore nozzles.

Example: How much water will be delivered from a 2½-inch discharge of a hydrant with a .85 coefficient of discharge and an 8-psi discharge pressure?

PUMP OPERATORS HANDBOOK

Step 1. Write the formula:

$$Q = 29.7 = d^2 \times \sqrt{P} \times c$$

Step 2. Determine the formula values:

d = 2.5 inches
P = 8 psi
c = .85

Step 3. Solve the equation:

$$Q = 29.7 \times (2.5)^2 \times \sqrt{8} \times .85$$
$$Q = 29.7 \times 6.25 \times 2.83 \times .85$$
$$Q = 446.5 \text{ gpm}$$

Example: How much water will be delivered from a 1¼-inch smooth bore nozzle if the nozzle pressure is 50 psi?

Step 1. Write the formula:

$$Q = 29.7 \times d^2 \times \sqrt{P} \times c$$

Step 2. Determine the formula values:

d = 1.25 inches
P = 50 psi
c = 1

Step 3. Solve the equation:

$$Q = 29.7 \times (1.25)^2 \times \sqrt{50} \times 1$$
$$Q = 29.7 \times 1.56 \times 7.07 \times 1$$
$$Q = 327.55 \text{ gpm}$$

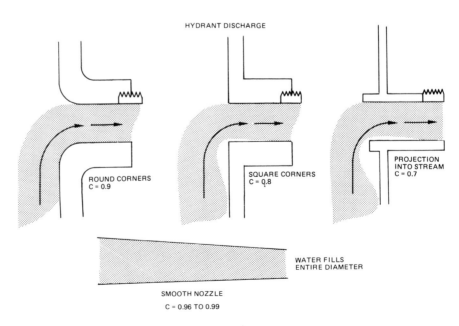

Figure 10. Coefficient of discharge.

Since the standard size nozzles available for the fire service are well known, the flow from these nozzles at various pressures can easily be calculated using the formula. Table 4 contains these calculations for nozzles from 1/8 to 3-inch diameters, at pressures from 20 to 150 psi.

As an example in using the table, find the flow from a 1¼-inch nozzle at 50 psi discharge pressure. Read down the 1¼ column and read across the 50 row. At the point where they intersect, read the value 326 gpm. Note that this value varies slightly from the flow calculated in the previous example. The discrepancy is due to the use of a coefficient of discharge for the nozzle, which reduces the value somewhat. The variation is less than .4 percent.

On fog nozzles, flow will also vary with nozzle pressure, but the nozzle pressure is difficult to measure. In addition, various manufacturers with differing models will have different flows for the same size nozzle at the same nozzle pressure. For this reason, when calculating using a fog nozzle certain standards can be set.

1½-inch nozzle — 100 psi nozzle pressure — 100 gpm
2½-inch nozzle — 100 psi nozzle pressure — 240 gpm
Master stream nozzle — 100 psi nozzle pressure — 500 to 1000 gpm

If the rated flow is known, then table 5 can be used to calculate the flows from that particular nozzle. For example, suppose a 2½-inch nozzle is rated at 240 gpm at 100 psi. What will this nozzle flow at a 70 psi nozzle pressure?

Locate the 240 gpm row in the left-hand column of table 5. Read across to the 70 psi column. At 70 psi, this nozzle will flow 201 gpm.

Simplified fireground flow calculations

The formula $Q = 29.7 \times d^2 \times \sqrt{P} \times c$ can be simplified to $Q = 30 \times d^2 \times \sqrt{P}$, with $c = 1$ for smooth bore nozzles. With the three nozzle pressures most used in the fire service (50, 80 and 100 psi), the calculations for the P can be simplified to $\sqrt{50} = 7$; $\sqrt{80} = 9$; $\sqrt{100} = 10$.

For a 1-inch nozzle at 50 psi nozzle pressure, this calculation becomes:

$$Q = 30 \times (1)^2 \times \sqrt{50}$$
$$Q = 30 \times 1 \times 7$$
$$Q = 210 \text{ gpm}$$

ENGINE PRESSURE CALCULATIONS

All of the material discussed so far in this chapter provides the background information necessary for calculating engine pressure. A pump operator must determine the proper engine pressure in order to deliver water in the most efficient manner possible. The basic equation for engine pressure is:

EP = NP + FL ± E, where
EP = engine pressure in psi
NP = nozzle pressure in psi
FL = friction loss in psi
+E = elevation in psi, where the discharge is above the pump
−E = elevation in psi where the discharge is below the pump

Example 1: What should the engine discharge pressure be to supply 400 gpm through a 200 foot 2½-inch line whose fog nozzle is located 60 feet above the pump?

PUMP OPERATORS HANDBOOK

Step 1. Select the proper equation:

$$E.P. = NP + FL + E$$

Step 2. Determine the formula values:

$$NP = \text{fog nozzle} = 100 \text{ psi}$$

$$FL\ (2\tfrac{1}{2}\text{-inch}) = 2Q^2 + Q$$
$$= 2(4)^2 + 4$$
$$= 36 \text{ psi}/100 \text{ ft}$$
$$= 36 \times \frac{200}{100} = 72 \text{ psi}/100 \text{ ft}$$

$$+E = 60 \times .434 = 26.04 \text{ psi}$$

TABLE 5. Discharge of Fog Nozzles, Gpm

Rated Flow of Nozzle at 100 psi (gpm)	Nozzle Pressure in Pounds per Sq In							
	30	40	50	60	70	80	90	100
	U.S. Gallons per Minute							
9	5	6	6.5	7	7.5	8	8.5	9
12	6.5	7.5	8.5	9	10	11	11.5	12
14	7.5	9	10	11	11.5	12	13	14
15	8	9.5	10.5	11.5	12.5	13	14	15
18	10	11	13	14	15	16	17	18
20	11	12.5	14	15	17	18	19	20
21	11.5	13	15	16	18	19	20	21
23	12.5	14.5	16	18	19	21	22	23
24	13	15	17	19	20	22	23	24
27	15	17	19	21	23	24	26	27
28	15	18	20	22	24	25	27	28
32	18	20	23	25	27	29	30	32
34	19	22	24	26	28	30	32	34
42	23	27	30	32	35	38	40	42
47	26	30	33	36	39	42	45	47
50	27	31	35	39	42	45	47	50
51	28	32	36	40	43	46	48	51
54	29	34	38	42	45	48	51	54
55	30	35	39	43	46	49	52	55
60	33	38	42	46	50	54	57	60
61	33	39	43	47	51	55	58	61
65	36	41	46	50	54	58	62	65
78	43	49	55	60	65	70	74	78
90	49	57	64	70	75	80	85	90
92	50	58	65	71	77	82	87	92
95	52	60	67	73	79	85	90	95
96	53	61	68	74	80	86	91	96
100	55	63	71	77	84	89	95	100
106	58	67	75	82	88	95	100	106
107	59	68	76	83	89	96	101	107
108	59	68	77	84	90	97	102	108
118	65	75	83	91	99	106	112	118
120	66	76	85	93	100	107	114	120
125	68	79	88	97	105	112	119	125
140	77	89	99	108	117	125	133	140
154	84	97	109	119	129	138	146	154
240	131	152	170	186	201	215	228	240
250	137	158	177	194	209	224	237	250
280	153	177	198	217	234	250	266	280
400	219	253	283	310	335	358	379	400
500	274	316	354	387	418	447	474	500
700	383	443	495	542	586	626	664	700
1000	548	632	707	775	837	894	949	1000
1350	739	854	955	1046	1129	1207	1281	1350
2000	1095	1265	1414	1549	1673	1789	1897	2000

TABLE 5. Discharge of Fog Nozzles, Gpm *(continued)*							
Rated Flow of Nozzle at 100 psi (gpm)	Nozzle Pressure in Pounds per Sq In						
	110	120	130	140	150	400	600
	U.S. Gallons per Minute						
9	9.5	10	10.2	10.5	11	18	22
12	12.5	13	13.5	14	14.5	24	29
14	14.5	15	16	16.5	17	28	—
15	16	16.5	17	18	18.5	30	—
18	19	20	20.5	21	22	36	—
20	21	22	23	24	24.5	40	—
21	22	23	24	25	26	42	—
23	24	25	26	27	28	46	—
24	25	26	27	28	29	48	—
27	28	30	31	32	33	54	—
28	29	31	32	33	34	56	—
32	34	35	36	38	39	64	—
34	36	37	39	40	42	—	—
42	44	46	48	50	51	—	—
47	49	51	53	56	57	—	—
50	52	55	57	59	61	—	—
51	53	56	58	60	62	—	—
54	57	59	61	64	66	—	—
55	58	60	63	65	67	—	—
60	63	66	68	71	73	—	—
61	64	67	69	72	75	—	—
65	68	71	74	77	80	—	—
78	82	85	89	92	95	—	—
90	94	99	103	106	110	—	—
92	96	101	105	109	113	—	—
95	100	104	108	112	116	—	—
96	101	105	109	113	117	—	—
100	105	110	114	118	122	—	—
106	111	116	121	125	130	—	—
107	112	117	122	127	131	—	—
108	113	118	123	128	132	—	—
118	124	129	135	140	144	—	—
120	126	131	137	142	147	—	—
125	131	137	142	148	153	—	—
140	147	153	160	166	171	—	—
154	161	169	175	182	189	—	—
240	252	263	274	284	294	—	—
250	262	274	285	296	306	—	—
280	294	307	319	331	343	—	—
400	420	438	456	473	490	—	—
500	524	548	570	592	612	—	—
700	734	767	798	828	857	—	—
1000	1049	1095	1140	1183	1225	—	—
1350	1416	1479	1539	1597	1653	—	—
2000	2098	2191	2280	2366	2449	—	—

Reprinted with permission from Akron Brass Company, Wooster, Ohio

Step 3. Solve the equation:

$$EP = 100 + 72 + 26 = 198 \text{ psi}$$

Example 2: What engine pressure is necessary to deliver 200 gpm through 200 feet of 2½-inch hose to a fog nozzle on level ground?

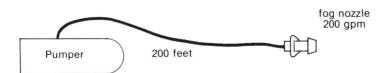

Step 1. Select the proper equation with a known flow:

$$EP = NP + FL + E$$

Step 2. Determine the formula values:

$$NP = \text{fog nozzle} = 100 \text{ psi}$$
$$FL = 2Q^2 + Q$$
$$= 2(2)^2 + 2$$
$$= 10 \text{ psi}/100 \text{ ft}$$
$$= 10 \times \frac{200}{100} = 20 \text{ psi}/200 \text{ ft}$$
$$E = 0 \text{ (level ground)}$$

Step 3. Solve the equation:

$$EP = 100 + 20 + 0$$
$$EP = 120 \text{ psi}$$

Example 3: What engine pressure is necessary to deliver 600 gpm to a deck gun with a fog nozzle, through two 3-inch lines over a 300-foot distance? The deck gun is 20 feet higher than the pumper.

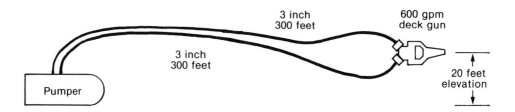

Step 1. Select the proper equation:

$$EP = NP + FL + E \text{ with flow being known}$$

Step 2. Determine the formula values:

$$NP = 100 \text{ psi}$$
$$FL_{2\frac{1}{2}} = 2Q^2 + Q$$
$$Q = 3 \text{ (each line carries 300 gpm)}$$
$$FL = 2(3)^2 + 3$$
$$FL_{2\frac{1}{2}} = 21 \text{ psi}/100 \text{ ft}$$
$$FL_3 = 21 \times .4 = 8.4 \text{ psi}/100 \text{ ft}$$
$$= 8.4 \times 3 = 25.2 \text{ psi}/300 \text{ ft}$$
$$FL \text{ (deck gun)} = 10 \text{ psi}$$
$$FL = 25.2 + 10 = 35.2 \text{ psi}$$
$$+E = 20 \times .434 = 8.68 \text{ psi}$$

Step 3. Solve the equation:

$$EP = NP + FL + E$$
$$EP = 100 + 35.2 + 8.68$$
$$EP = 144 \text{ psi}$$

Example 4: What engine pressure is necessary to deliver 700 gpm to a deck gun with a 1-5/8-inch tip through 200 feet of 2-1/2-inch and 200 feet of 3-inch parallel lines? The deck gun is 10 feet below the pumper.

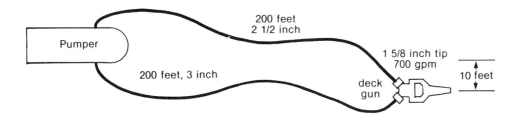

Step 1. Select the proper equation, with flow being known:

$$EP = NP + FL - E$$

Step 2. Determine the formula values:

$$NP = \text{straight tip, master stream} = 80 \text{ psi}$$
$$FL_{2\frac{1}{2}} = 2Q^2 + Q$$
$$FL_{2\frac{1}{2}} = 2(7)^2 + 7$$
$$FL_{2\frac{1}{2}} = 105 \text{ psi}/100 \text{ ft}$$
$$FL_{2\frac{1}{2} + 3} = 105 \times 1/6$$
$$= 17.5 \text{ psi}/100 \text{ ft}$$
$$FL_{2\frac{1}{2} + 3} = 17.5 \times \frac{200}{100}$$
$$= 35 \text{ psi}/200 \text{ ft}$$
$$FL_{DG} = 10 \text{ psi}$$
$$FL = 35 + 10 = 45 \text{ psi}$$
$$-E = -10 \times .434 = -4.34$$

Step 3. Solve the equation:

$$EP = 80 + 45 - 4$$
$$EP = 121 \text{ psi}$$

Example 5: What engine pressure is necessary to supply the setup shown in the illustration:

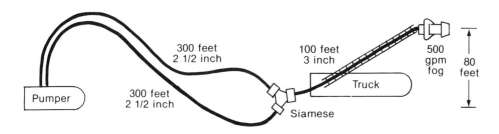

Step 1. Select the proper equation with flow being known:

$$EP = NP + FL + E$$

Step 2. Determine the formula values:

$$NP = 100 \text{ psi for a fog nozzle}$$

From siamese to nozzle

$$\begin{aligned} FL_{2\frac{1}{2}} &= 2Q^2 + Q \\ &= 2(5)^2 + 5 \\ &= 55 \text{ psi}/100 \text{ ft} \\ FL_3 &= 55 \times .4 = 22 \text{ psi}/100 \text{ ft} \end{aligned}$$

From pumper to siamese

$$Q = \frac{500}{2} = 250 \text{ gpm in each line}$$

$$\begin{aligned} FL_{2\frac{1}{2}} &= 2(2.5)^2 + 2.5 \\ &= 15 \text{ psi}/100 \text{ ft} \\ &= 15 \times \frac{300}{100} = 45 \text{ psi}/300 \text{ ft} \end{aligned}$$

$$\begin{aligned} FL_{siam} &= 5 \text{ psi} \\ FL_{LP} &= 10 \text{ psi} \\ FL &= 22 + 45 + 5 + 10 = 82 \text{ psi} \\ +E &= 80 \times .434 = 34.72 \end{aligned}$$

Step 3. Solve the equation:

$$EP = 100 + 82 + 35 = 217 \text{ psi}$$

Example: What engine pressure is necessary to supply the setup shown in the illustration?

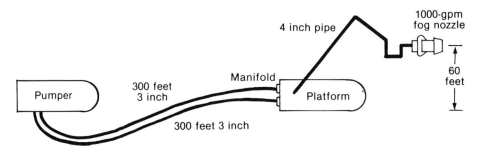

Step 1. Select the proper equation, with flow being known:

$$EP = NP + FL + E$$

Step 2. Determine the formula values:

$$NP = \text{fog nozzle} = 100 \text{ psi}$$
$$FL_{pipe} = 21.8 \text{ (table 3)}$$

FIREGROUND HYDRAULICS

$$Q = \frac{1000}{2} = 500 \text{ gpm}$$

$$\begin{aligned}
FL_{2\frac{1}{2}} &= 2Q^2 + Q \\
&= 2(5)^2 + 5 \\
&= 55 \text{ psi}/100 \text{ ft}
\end{aligned}$$

$$FL_3 = 55 \times .4 = 22 \text{ psi}/100 \text{ ft}$$

$$FL_3 = 22 \times \frac{300}{100} = 66 \text{ psi}/300 \text{ ft}$$

$$FL_{LP} = 10 \text{ psi}$$

$$FL_{mani} = 5 \text{ psi}$$

$$FL = 66 + 10 + 5 + 22 = 103 \text{ psi}$$

$$+E = 60 \times .434 = 26 \text{ psi}$$

Step 3. Solve the equation.

$$EP = 100 + 103 + 26 = 229 \text{ psi}$$

Example 7: What engine pressure is necessary to supply to 200-foot 2½-inch hand lines, supplied from a 300-foot length of 3-inch line, with a 1-inch nozzle tip at 50 psi nozzle pressure?

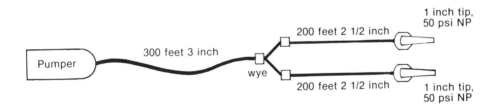

Step 1. Select the proper equation, when flow is unknown:

$$Q = 29.7d^2 \sqrt{P} \, c$$

After flow is calculated:

$$EP = NP + FL + E$$

Step 2. Determine the formula values:

$$\begin{aligned}
Q &= 29.7 \times (1)^2 \times \sqrt{50} \times 1 \\
&= 29.7 \times 7.07 \\
&= 210 \text{ gpm in each 2½-inch line}
\end{aligned}$$

$$NP = 50 \text{ psi, straight tip, hand line}$$

From wye to nozzle:

$$\begin{aligned}
FL_{2\frac{1}{2}} &= 2(2.1)^2 + 2.1 \\
&= 10.92 \text{ psi}/100 \text{ ft} \\
&= 11 \times \frac{200}{100} = 22 \text{ psi}/200 \text{ ft}
\end{aligned}$$

$$FL_{wye} = 5 \text{ psi}$$

From pumper to wye:

$FL_{2½} = 2Q^2 + Q$

$Q = 210 + 210 = 420$ gpm

$FL_{2½} = 2(4.2)^2 + 4.2$

$= 39.48 = 39$ psi/100 ft

$FL_3 = 39 \times .4 = 15.6 = 16$ psi/100 ft

$FL_3 = 16 \times \dfrac{300}{100} = 48$ psi/300 ft

$E = 0$

$FL = 22 + 5 + 48 = 75$ psi

Step 3. Solve the equation:

$EP = 50 + 75 + 0 = 125$ psi

Chapter 7

Introduction to Pump Operation

The progress in the development of pumps was closely tied to the progress made in the field of hydraulics. Limitations caused by lack of sufficient, easily accessible water supplies created the need for a mechanical device that could move water from one location to another for firefighting use. One of the earliest mechanical devices, or pumps, was developed about 200 B.C. by Ctesibius. This pump consisted of two brass cylinders with carefully fitted pistons that drew water through valves at the base and discharged it through outlet valves into a chamber.

In other areas of the world, even cruder devices were being used as pumps. Marco Polo reported that professional firefighters in Cathay used hand-operated siphons as pumps and crude hose fashioned from oxen intestines about 1300 A.D.

Leather hose with brass fittings was first made in Holland in 1672. This hose was made from the finest cowhide, with the seams carefully sewn together. Sleeves, for drafting water from the source to the pump, were made of heavy sailcloth with stiff paint, reinforced with metal rings to prevent them from collapsing when subjected to a negative pressure by the pump. A brass strainer was lowered into the water source. This helped to keep the hose and pump free from dirt.

Firefighting in this country began with the Dutch in New Amsterdam (New York). In the early 1600s, they established a night fire watch and required that each household have a leather bucket for use in fighting a fire. In 1679, Boston

Figure 1. A 1731 Newsham hand pumper from New York City.

purchased the first fire engine — a small, hand-drawn, hand pumper. In 1731, New York City received its first two hand pumpers, built in England by Richard Newsham (figure 1). With 10 men operating on each side, this pumper could deliver enough water for a single stream. The pumper had a small built-in reservoir called a cistern, that had a capacity of 170 gallons. However, since the pumper didn't have drafting sleeves, the engine could not lift water into the cistern, and water had to be supplied by buckets carried by hand. To speed up the process of filling the cistern, a multiple bucket carrier was developed (figure 2).

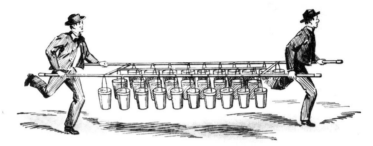

Figure 2. A multiple bucket carrier.

While hose had been developed in Europe, it had not yet come into use in America. Lines did not exist that could be brought into a building to fight fires. A stream of water was directed from the nozzle or pipe which was firmly attached to the pumper.

In the 1820s, the gooseneck engine evolved as a distinctly American-style engine. It was large and bulky and required 12 men to pull it to fires. Hose was developed that could be attached to this apparatus so that rarely operations from the water source to the fire could be set up. Also, this unit carried sleeves for drafting water into the pumper as well as hose which could be attached to it. Firefighters were now able to move the nozzle from the engine and stretch it closer to the fire. The hose that went from the engine to the fire was called the discharge hose. It was made of leather, came in 50-foot lengths with brass couplings, and weighed about 90 pounds per length.

Although rubber-treated hose was introduced in 1827, and woven-cotton and linen hose was used prior to 1827, these more practical materials did not begin to replace leather hose until late in the 19th century. In some of the smaller fire departments, the use of leather hose persisted until the 1900s.

As stated before, a pumper was only as good as the water supply available. While technology indicated that larger pumps, with greater capacity, could be built, the development of suitable water systems did not grow at the same pace.

One of the early water supply conduits was developed in Boston in the early 1800s to bring water from the outskirts to the downtown area in order to supply water for fire department use. The system consisted of a 15-mile length of pitch pine logs with 3 or 4-inch bores. Every half block, a wooden plug (origin of the term "fire plug" for hydrant) was inserted into the pipe. This plug could be removed in case of fire and water could be withdrawn from the log at this point. Unfortunately, the flow was so weak that very little water could be withdrawn. Since this system was unsatisfactory, water had to be taken from wells and cisterns; reservoirs were strategically located and supplied from these wells sunk in the vicinity. Of course, this method required long relays of pumping engines to bring the water from the source to the fireground. This was a time-consuming procedure to set up, especially because of the time it took for many of the pumpers to arrive.

Relays, of course, were limited in length by the amount of hose carried on the apparatus. Additional hose was carried to the fire on the shoulders of the firefighters, but the 90-pound hose made this method ineffective. The solution to this problem was the development of a two-wheel, box-like cart, that carried 50-foot, coiled lengths of hose. When fire companies demanded still more hose, the two-wheel hose reel known as a jumper, leader cart, or tender was developed. This device carried 600 feet of hose and could be hand-drawn, or pulled by attaching it to the tail hook of the engine (figure 3).

Figure 3. A two-wheeled hose reel.

Figure 4. Hose carriage belonging to Weiner Hose Company 6, Kingston, N.Y.

Separate hose companies were formed because the number of people needed to lay hose could not be spared from the engine companies. The hose companies, desiring distinctive apparatus, developed the four-wheel hose carriage that usually carried 1000 feet of hose and weighed between 1200 and 1500 pounds (figure 4). To wind the hose on the drum, a winch handle operated a series of gears that turned the drum.

Many different types of hand-operated pumpers were developed during the 1800s. The Philadelphia-style engine was an end-stroke type with the pumping mechanism located in the front and back of the apparatus. Some of these engines had two sets of pumping handles (double-decker brakes), so that more water could be pumped (figure 5).

Another type of hand pumper, which appeared about 1850, was the piano or squirrel tail pumper. This apparatus had a permanently attached drafting hose which was swung up over the top of the engine when not in use and secured by a large brass pipe. The walking beams or cross arms of the pumping device were slotted so that leverage could be altered without changing the depth of stroke. The pumping handles (brakes) could also be folded when not in use.

Figure 5. A 1853 Philadelphia-style hand pumper.

Figure 6. First steamer in the United States, 1853, Cincinnati, Ohio.

The first steam fire engine was built in 1829 in London. When the steam pumper was introduced, the volunteer firefighters refused to accept or use it because they felt it threatened their firefighting jobs. They even went as far as to refuse to supply water to any steamer on the fireground. The steamer, however, did have arithmetic working in its favor. For example, it could take the place of 12 hand pumpers, with a resulting manpower saving of from 50 to 75 men per pumper. It was obvious that this manpower savings would cause many municipal governments to favor the steam pumper over the hand pumper.

The City of Cincinnati put the first steamer in the United States in operation in 1853 (figure 6). Weighing over 22,000 pounds and propelled by steam with the aid of four horses, it was also able to pump six hose lines at the same time, a feat that was impossible with a hand machine. Even though steel and brass began to replace the cast iron construction of the steamer so that it was lighter, horses still were needed to pull the apparatus.

Early steam fire engines, like their predecessors the hand pumpers, were piston-type pumpers. However, as the steam era progressed, rotary-type pumpers gradually replaced the piston types.

While firefighters were reluctant to accept the horse-drawn steamer, they were even more resistant to accepting gasoline engines. In 1906, Wayne, Pa., acquired a gasoline-driven engine. Other attempts at gasoline power involved replacement of the horses with a completely self-contained, two-wheeled engine. This two-wheel tractor could be dismounted from the steamer and replaced with another tractor whenever the motor needed repairs (and that

Figure 7. Gasoline engine drives both pump and engine.

was frequently). By 1910, gasoline engines that would both drive the apparatus and operate the pump were developed (figure 7).

The development of the gasoline engine was paralleled by the progress made with the centrifugal pump. This was a major step because the gasoline engine could drive the centrifugal pump at high revolutions per minute, thus permitting greater discharge, at higher pressures, for a more efficient operation.

Pumpers have progressed to the point where they can deliver 2000 gallons per minute, carry 1000 gallons of water in a booster tank, and have 2000 feet of 3-inch hose (figure 8). The Super Pump System of the New York City Fire Department can pump 8800 gpm at a pressure of 350 pounds per square inch. This means that the pumper could supply 30 hose lines or from 10 to 20 small-bore deck guns with a total of more than 37 tons of water a minute (figure 9).

The fire service has come a long way since Ctesibius invented the piston pump. From the hand pumps which required 50 men to operate, pumping 60 to 120 strokes per minute, to the steamer with its rotary pump, the fire service pumpers have evolved to modern apparatus and centrifugal pumps. Industrial development, high-rise buildings, suburban sprawl, new materials, and large crowded cities will continue to force technological advances in the field of pumps and pumpers. As the future unfolds, many new innovations will be introduced.

Figure 8. 1969 Barton American pumper in the Berlin, Md., Fire Department.

Figure 9. Overview of the New York City Fire Department's Super Pumper.

Chapter 8

Positive Displacement Pumps

With the field of pumps and pumpers expanding, and new innovations and technologies being developed, "The fire apparatus driver/operator shall demonstrate how to operate the different types of fire department pumpers used by the authority having jurisdiction."*

One type of pump used by the fire service is positive displacement, which includes piston and rotary pumps. One of the rules of hydraulics states that pressure from the outside, when applied to water in a confined area, will be distributed to all parts of the area without decreasing in value. It is on the basis of this rule, the fact that water is incompressible, that a positive displacement pump works.

Positive displacement means that the volume of space within the pump will be the amount of water that the pump can deliver on one stroke or revolution. A mechanical method of increasing the volume, causing a pressure drop, brings the water into the pump. Then, when the volume is decreased, the increasing pressure forces the water out. The two major types of positive displacement pumps are piston and rotary pumps.

PISTON PUMPS

As discussed in the previous chapter, water pumps were known thousands of years ago. These early pumps were positive displacement pumps, more particularly, piston pumps (figure 1A).

Piston pumps for firefighting today are being used for high-pressure applications (figure 1B) and on apparatus for combatting brush and woods fires.

The quantity of water that can be delivered from a piston pump is determined by:
1. The size of the piston,
2. The length of the stroke,
3. The number of strokes per minute (rpm), and
4. The number of cylinders.

Lift pump

The most basic type of piston pump is the lift pump. The lift pump will only discharge water under a small amount of pressure. The major components of

Paragraph 3-6.4. Reprinted with permission from NFPA 1002-1982, Standard for Fire Apparatus Driver/Operator Professional Qualifications, Copyright© 1982, National Fire Protection Association, Quincy, Massachusetts 02269. This reprinted material is not the complete and official position of the NFPA on the referenced subject, which is represented only by the standard in its entirety.

POSITIVE DISPLACEMENT PUMPS

Figure 1A. Early piston pumper.

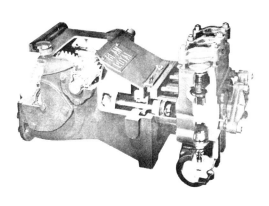
Figure 1B. John Bean high-pressure piston pump.

Figure 1C. Ahrens-Fox piston pumper.

the lift pump (figure 2A) are: the inlet to bring water into the lower chamber; the inlet valve for allowing water into the pump cylinder; the piston for moving the water; the piston valve for allowing the water to pass into the upper chamber; and the discharge for allowing water to leave the upper chamber. The pump operates as follows:

Step 1. The piston begins operation at the bottom of the pump cylinder. With no movement of the piston, both the inlet and piston valves are closed (figure 2B). At this time, the pump cylinder is filled with air, the lower chamber has no volume, and the upper chamber has all the volume.

Step 2. Now the piston is moved upward (figure 2C). As the piston moves upward, the volume in the lower chamber increases. This results in a decrease in pressure. Atmospheric pressure on the water is greater than the internal pressure of the lower chamber and water is pushed up into the lower chamber, opening the inlet valve. In the upper chamber, the rapid decrease in volume causes a temporary increase in pressure (temporary because the discharge is open to the atmosphere), which keeps the piston valves closed. At the maximum height of the piston, the lower chamber is filled with water.

Step 3. When the piston starts down again, the increased pressure in the lower chamber closes the inlet valve (figure 2D). Since water is incompressible, it goes into the upper chamber by opening the piston valve.

Step 4. The piston now starts its upward movement again (figure 2E). The lower chamber volume increases, pressure decreases, the inlet valve opens, and water enters the lower chamber. The water in the upper chamber, being incompressible, has no place to go as the volume is made smaller. The piston valve closes and the water is forced or lifted out the discharge.

Step 5. The procedure is repeated with water being discharged with each upward movement of the piston.

This pump is a positive displacement pump because the amount of water

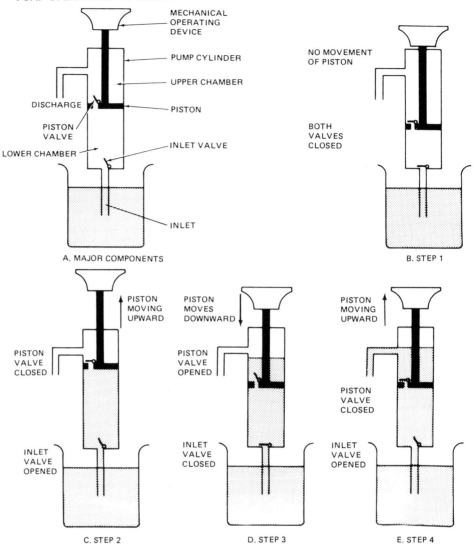

Figure 2. Lift pump.

pumped is directly dependent upon the volume of water that the pump cylinder will hold. (It is also dependent upon other factors which will be discussed later in this chapter.) This type of pump is also known as a *single-acting pump* because water is pumped only on one strike (1/2 stroke cycle).

Note that the pump starts by being filled with air and then the air is replaced by water. No other devices are necessary to remove the air from the pump. All positive displacement pumps are, therefore, considered self-priming.

Pressure pump

The pressure piston pump is a device that uses the piston to force water out the discharge. By regulating the amount of force applied by the piston, the pressure under which the water is discharged can also be regulated (figure 3A). This pump operates as follows:

Step 1. As the piston begins its upward movement, the increased volume of the lower chamber causes reduced pressure (figure 3B). Water is then pushed upward into the lower chamber. The lower pressure inside the lower chamber also keeps the discharge valve closed.

Step 2. When the piston moves downward, the increased pressure closes the inlet valve and opens the discharge valve (figure 3C). Water is then forced out the discharge under pressure. The amount of pressure at the discharge opening depends on the size of the opening.

POSITIVE DISPLACEMENT PUMPS

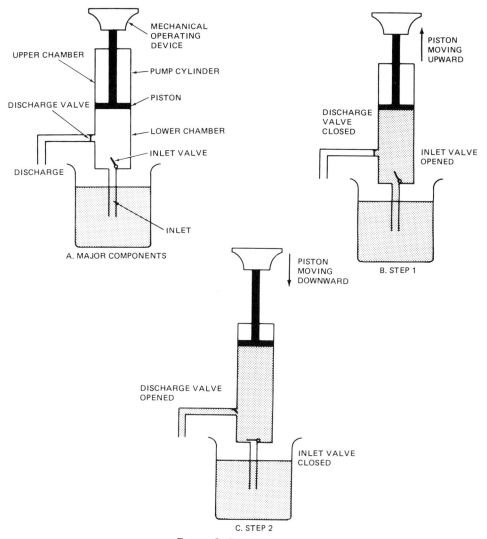

Figure 3. Pressure pump.

Step 3. The procedure is repeated with water being discharged with each downward movement of the piston.

This pump is also a positive displacement, single-acting and self-priming pump.

Both the lift pump and the pressure pump will deliver water only during one stroke. This means that the firefighter on one end of the line would experience a spurt of water and then nothing while the piston was on its return stroke. This is not acceptable for firefighting.

Air chamber for piston pumps

To overcome this difficulty, an air chamber is connected to the discharge side of the pump. Figure 4 shows the air dome on an Ahrens-Fox piston pumper. The air chamber (figure 5A) operates as follows:

Step 1. As the piston moves upward, the increased volume of the lower chamber causes reduced pressure (figure 5B). Water enters the lower chamber as the discharge valve slides shut.

Step 2. When the piston moves downward, the increased pressure closes the inlet valve and slides open the discharge valve (figure 5C). Water is then forced out the discharge piping and into the air chamber. Since the nozzle diameter is much smaller than the discharge piping diameter, water that cannot

PUMP OPERATORS HANDBOOK

Figure 4. Piston pump air chamber.

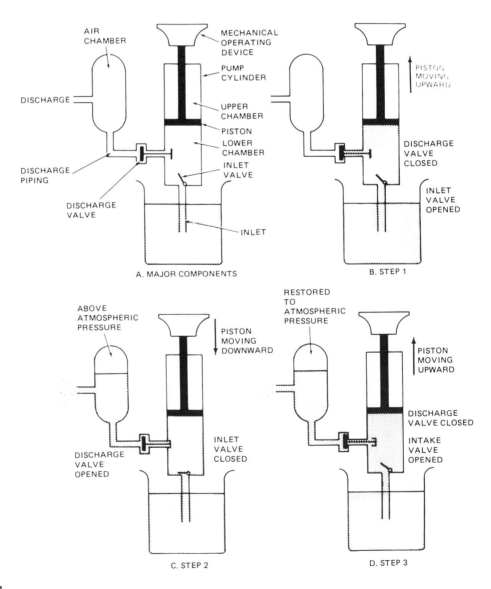

Figure 5. Operation of a piston pump with air chamber.

POSITIVE DISPLACEMENT PUMPS

be discharged out the nozzle is forced into the air chamber. The air within the chamber is now compressed by the water increasing the air pressure above atmospheric.

Step 3. Now the piston begins its upward movement again, and there should be no discharge (figure 5D). However, with the discharge valve closed, the air pressure in the air chamber causes the water to flow out the nozzle under pressure. Thus, on the nonpumping stroke, water continues to flow, smoothing out the pump discharge.

Step 4. The procedure is repeated, with water being discharged during both strokes of the piston.

It should be remembered that the flow with this type of pump is still not very smooth. A comparison of what the discharge might look like is shown in figure 6A.

The piston pump without the air chamber will have a period without any water being delivered during the return stroke (figure 6B); with an air chamber, this gap is filled in with water being supplied. Note, however, that a smooth, constant amount of water is still not being delivered.

The size of the air chamber is determined by the amount of water to be pumped. The larger the pump, the larger the air chamber that is needed.

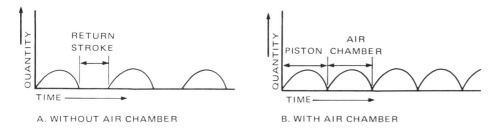

Figure 6. Discharge of piston pumpers.

Double-acting piston pump

One way to smooth out the delivery of a piston pump (and also increase its output) is to use a double-acting pump. A double-acting pump takes in water and pumps on both the forward and return strokes. In order to accomplish this, a dual set of intake and discharge valves are necessary (figure 7A). The pump operates as follows:

Step 1. As the piston moves upward (figure 7B), the volume increases, the pressure decreases, a the water from the intake opens intake valve 1 (IV 1). Water that is already in the piston's upper chamber forces IV 2 to close and opens discharge valve 2 (DV2).

Step 2. As the piston starts its downward movement, (figure 7C), IV 1 and DV2 are forced closed. IV 2 opens to allow water into the upper chamber and DV1 opens to allow water to be discharged.

Step 3. The procedure is repeated with water being discharged during both strokes of the piston.

Now, if an air chamber were added to a double-acting piston pump, the discharge might appear as shown in figure 8. The water delivery is still somewhat uneven.

Multiple-cylinder piston pumps

The final method for smoothing out the discharge is to use multiple cylinders, each discharging at a slightly different time interval. Both single-acting and double-acting piston pumps can be used with multiple cylinders. Single-acting pumps are usually arranged so that they change directions at

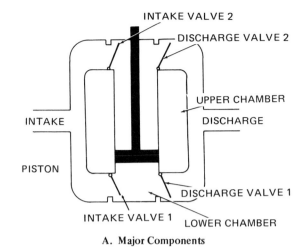

A. Major Components

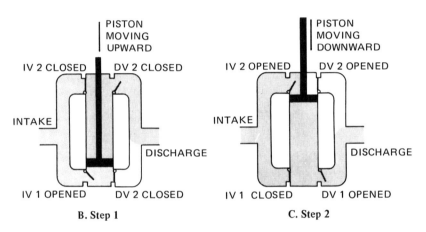

B. Step 1   C. Step 2

Figure 7. Double-acting piston pump.

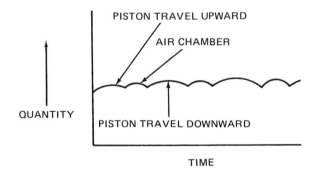

Figure 8. Discharge of a double-acting piston pump with air chamber.

staggered intervals. This not only smooths out the water delivery curve, but helps to balance out the pump.

Double-acting pumps, with multiple cylinders, must be arranged in pairs. Again, the change in direction of travel is staggered so that a smooth discharge is achieved and balance of the pump is maintained.

Slippage

In the construction of piston pumpers, it is extremely important that very close tolerances be used between the wall of the cylinder and the piston. These close tolerances are necessary to prevent water from slipping back to the intake

POSITIVE DISPLACEMENT PUMPS

side of the piston from the discharge side. However, because some space is necessary, a small amount of water does get back. This backward movement from discharge to intake is called slippage. As the piston wears, the amount of slippage increases until the piston head has to be replaced.

ROTARY PUMPS

Rotary pumps for firefighting were developed long after the piston pump. Although they reached their peak of use as main fire pumps during the era of the steamer, they were adapted to the gasoline-powered engine. While their use as a main fire pump has declined, their self-priming ability makes them useful on modern-day apparatus as priming pumps. In addition, rotary pumps make excellent auxiliary pumps.

The quantity of water that can be delivered from rotary pump is determined by:
1. The amount of space between gear teeth,
2. The number of revolutions per minute.

While the number of pistons used in a piston pump affects the quantity delivered, increasing the number of teeth of a rotary pump does not produce more water. Of sole importance on a rotary pump is the amount of space between the teeth.

Rotary pumps can be divided into three general categories: rotary gear, clover leaf, and eccentric rotary vane.

Rotary gear and clover leaf

Rotary gear and clover leaf types of rotary pumps operate in the same manner, except for a differing number of gear teeth. The rotary gear design usually has eight gear teeth (figure 9), although other amounts can be found; the clover leaf always has three teeth (figure 10). These pumps operate as follows:

Step 1. Water or air enters the intake and is picked up between the teeth of the gear. Either of the gears may be driven (rotated by a power source) or sometimes both gears can be driven.

Step 2. As the teeth rotate, the water is moved toward the discharge opening.

Step 3. As the space between teeth reaches the discharge opening, water or air is forced out. The water or air has no place to go due to the meshing of the gear teeth, and is, therefore, forced out the discharge. Pressure is built up as

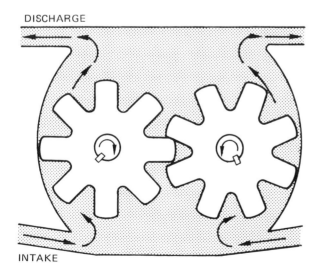

Figure 9. Gasoline-powered rotary gear pump.

PUMP OPERATORS HANDBOOK

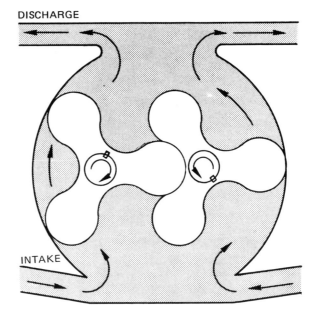

Figure 10. Clover leaf pump.

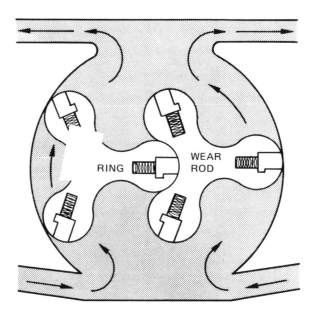

Figure 11. Clover leaf pump with wear rod.

the rapidly turning teeth force more water out the discharge. The meshing of the gear teeth also prevents water from returning to the intake side of the pump.

As in the case of the piston pump, slippage occurs, with one of the biggest causes being wear. To overcome this deficiency, a clover leaf pump with wear rods was designed (figure 11). The rod is kept against the housing by a spring to provide a better seal and to reduce slippage. As the rod wears down, the spring pushes it out further to continue to seal the housing. When the wear has exceeded the available limits, just the wear rod must be replaced rather than the whole clover leaf assembly.

POSITIVE DISPLACEMENT PUMPS

Rotary vane

To operate, the rotary vane pump uses only a single rotor, which is off-center (figure 12). The pump operates as follows:

Step 1. Water or air enters the intake because the increasing space between the vanes causes a pressure drop. The vanes are kept against the housing by

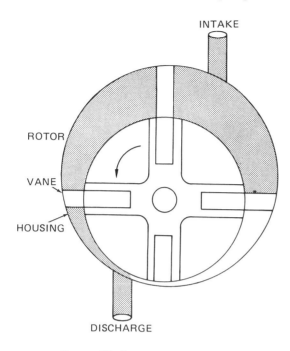

Figure 12. Rotary vane pump.

centrifugal force and sometimes a spring is used to assist in keeping the vanes in place.

Step 2. As the vanes rotate, the water is moved toward the discharge opening.

Step 3. As the water or air nears the discharge opening, the space narrows because the rotor is off center. The water or air, having no place to go, is forced out the discharge, thus building up pressure.

As with the other rotary pumps, slippage occurs due to water being bypassed around to the intake side of the pump.

Air chamber for rotary pumps

Rotary-type pumps deliver a fairly steady stream of water. This is accomplished by gearing the teeth so that as one space finishes discharging, the opposite gear has a space that is just beginning to discharge. However, as pump size and flows increase, 750 gpm or higher, an air chamber is needed. This air chamber does not have to be quite as large as that for a piston pumper, but it operates in the same manner.

Chapter 9

Centrifugal Pumps

While a positive displacement pump discharges a definite volume of water for each cycle, a non-positive displacement pump discharges a volume of water based upon its inherent resistance to movement. The force exerted on the water depends on the speed at which the pump is operating.

One type of non-positive displacement pump operates on the theory of a rapidly spinning disk to create a force known as centrifugal force. This is the most common type of pump used in the fire service and is known as a centrifugal pump.

"The fire apparatus driver/operator shall identify the operating principles of single stage and multiple stage centrifugal fire pumps.

"The fire apparatus driver/operator, given pump models or diagrams, shall identify the major components and trace the flow of water through single stage and multiple stage centrifugal pumps.

"The fire apparatus driver/operator, given a fire department pumper with a multiple stage pump, shall demonstrate the use of the volume/pressure transfer valve under actual pumping conditions."[*]

While the principles of operation for all centrifugal pumps are the same, each major manufacturer varies the method of moving the water. After a general discussion applicable to all pumps, we will cover the individual differences of the major manufacturers.

Centrifugal force

If a small amount of water were placed at the center of a rotating disk, the water would be thrown outward (figure 1). The rotating effect moves the water from a standing position by imparting a horizontal velocity. If the speed of rotation were increased, the water would be thrown farther. The force that makes rotating bodies move away from the center of rotation is known as centrifugal force.

Another way of demonstrating centrifugal force is to attach a string to a paper cup half full of water. Begin to swing the string and cup in an ever-increasing arc until finally completing a full circle. The water remains within the cup due to centrifugal force. This force applies a pressure on the surface of the liquid in the cup much greater than the force of gravity and atmospheric pressure combined.

---

[*]Paragraphs 3-1.2, 3-1.3 and 3-6.7. Reprinted with permission from NFPA 1002-1982. Standard for Fire Apparatus Driver/Operator Professional Qualifications, Copyright©1982, National Fire Protection Association, Quincy, Massachusetts 02269. This reprinted material is not the complete and official position of the NFPA on the referenced subject, which is represented only by the standard in its entirety.

CENTRIFUGAL PUMPS

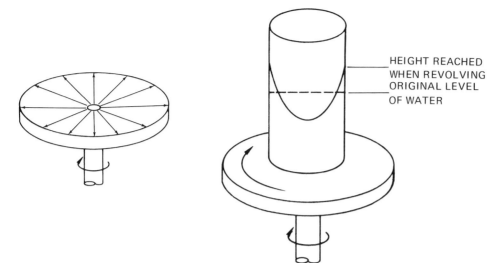

Figure 1. Centrifugal force from a rotating disk.

Figure 2. Conversion of centrifugal force from velocity to pressure.

Now to show the conversion of velocity to pressure, place a glass of water on a revolving disk (figure 2). As the disk turns, the water will be thrown to the sides, but because it is confined within the walls, the level will rise. In the previous chapter, height of water was shown to be directly related to pressure head. So, by confining the liquid as it is rotated, the centrifugal force can be changed to pressure head. The height of the liquid within the container depends on the speed of rotation.

The three factors that regulate the effectiveness of a centrifugal pump — pressure, speed and quantity — are all interrelated. If pressure is kept constant, then an increase in speed will increase the quantity. If the quantity is kept constant, an increase in speed will cause an increase in pressure. With speed kept constant, an increase in pressure will cause a decrease in quantity:

$$\frac{speed}{pressure} = quantity$$

Pump components

The major components of a centrifugal pump (shown in figures 3 and 4) are:

Impeller — The impeller is the major component of the centrifugal pump, because it provides the velocity to the water. The impeller is mounted on a shaft which is rotated by some type of motor power (figures 3, 4 and 5). (The various methods of driving the impeller are covered in Chapter 10.)

Water enters the rapidly revolving impeller at the intake or eye (figures 4 and 5), and is confined by the sides, called *shrouds,* and by the *vanes* inside the impeller. As explained, confining the water forces it toward the outer edge at increased pressure.

Additional pressure is created because the outer edge of the impeller is rotating faster than the eye due to the increased diameter at the edge. This means that water travels a greater distance near the discharge than at the intake, and a greater pressure results.

The vanes guide the water from the inlet to the discharge and minimize the turbulent effect that spinning water produces. The vanes are curved away from the direction of rotation so that the natural movement of the water will carry it to the edge.

Figure 6 traces a drop of water from the eye of the impeller to the discharge outlet. Note that in step 7 of the figure, just as the drop reaches the outer edge,

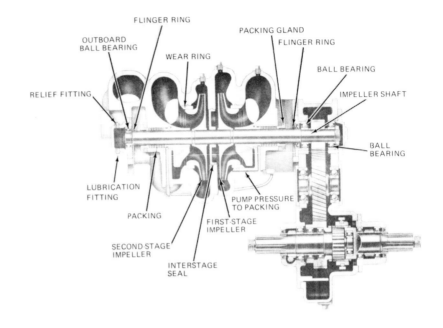

Figure 3. Cross section of a centrifugal pump.

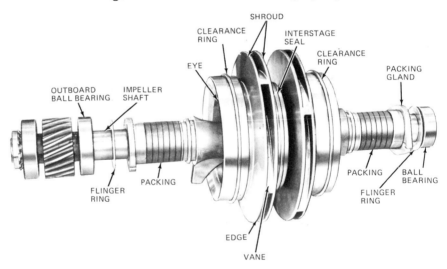

Figure 4. Impeller shaft components.

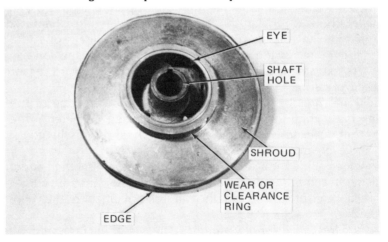

Figure 5. Impeller components.

CENTRIFUGAL PUMPS

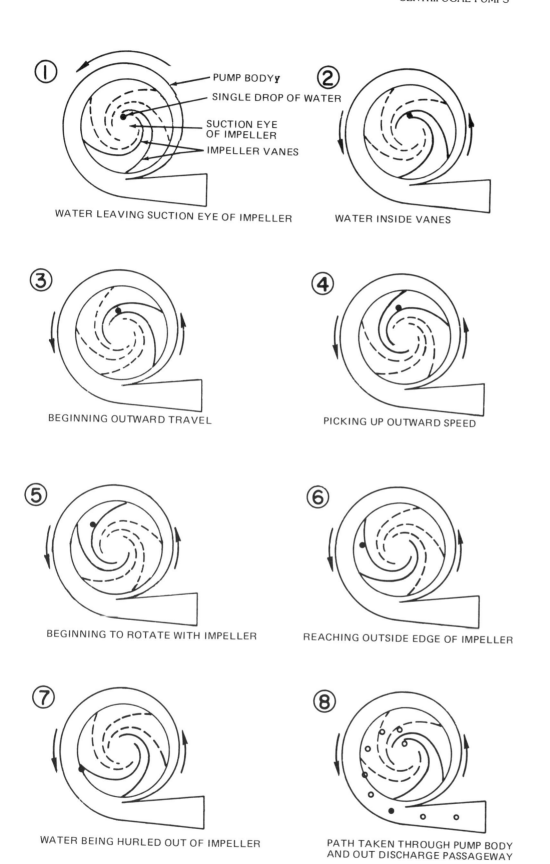

Figure 6. Operation of a centrifugal pump.

93

the maximum opening of the vane is at the discharge. The impeller is also positioned off-center so that the discharging tube is at its widest at the pump outlet. The gradually increasing discharge allows a gradual shift by reducing velocity and increasing pressure. The gradually increasing discharge is known as the *volute*.

Impellers can be either single-suction or double-suction, depending on the flows and pressures required. On the single-suction type, water enters on the one side with the eye, while a double-suction impeller has two eyes, with water entering from both sides.

One of the factors that affects the capacity of the pump is the size of the impellers. However, the pump is rated from a draft condition, which means a negative pressure at the eye of the impeller. When operating from a pressure source such as a hydrant, the pump can discharge greater quantities of water than its rated capacity. One of the major advantages of centrifugal pumps over the positive displacement pump is that it takes advantage of incoming pressure. For example, if hydrant pressure to the pump is measured at 50 psi, and the centrifugal pump is rated at 150 psi, it is then able to apply the 150 pounds to the incoming supply at 50 pounds and discharge a maximum of 200 psi. A high volume of water under sufficient pressure may allow the pump to deliver up to 200 percent of its rated capacity.

Wear or clearance rings — As the water leaves the discharge, it is necessary to prevent it from returning to the eye of the impeller. Centrifugal pumps use a wear or clearance ring at the eye of each impeller to prevent this leakage (called slippage in a positive displacement pump). (See figures 3, 4 and 5.)

This ring usually has a clearance of about .006 inch. This small opening will increase as the pump is operated, and the ring is designed for replacement. However, if the pump should be run without water or without discharging water, the pump will overheat, the rings will expand, and damage to the pump and the drive mechanism can occur.

Bearings — The bearings provide support and alignment for the impeller shaft so that a smooth rotation under the dynamic stress of water flow can occur (figures 4 and 5).

Packing — Pump packing is the device that allows the impeller shaft to pass from the outside of the pump to the inside while maintaining an airtight seal (figures 3 and 4). At each entrance or exit of the shaft, packing and a packing gland are installed. The packing material is designed to be lubricated by pump water. The packing gland should, therefore, not be tightened too tightly for the packing material will dry out, crumble, and cause pump damage. Another way of causing the packing material to dry out is to keep the pump dry for long periods of time during the winter months. If the pump is kept dry to prevent freezing, then at least once a week the pump should be charged with water to lubricate the packing.

Flinger ring — Since water is designed to keep the packing material wet, a method is needed to keep the water from continuing to travel the impeller shaft to the gears and ball bearings. This is accomplished by a flinger ring, which throws water off of the impeller shaft.

Stages — The number of impellers mounted on a common shaft determine the number of stages of a pump. The additional stages allow some versatility in operating the pump at various pressures and flow, while keeping the drive motor operating in a manageable range.

Single-stage pumps have been developed that will deliver the same range of volume and pressures as a two-stage pump. Improvement in engine design and ease of operation (no transfer valve) make the single-stage pump a popular design.

CENTRIFUGAL PUMPS

Figure 4 shows a two-stage pump with two impellers mounted on the same shaft. Figure 7 is a cut-away view of a four-stage pump.

The two-stage pump can handle large quantities of water at low pressures, or small quantities of water at high pressures at approximately the same engine speed.

Just as clearance rings are needed at the eye of each impeller, an interstage seal is necessary to prevent water from the discharge of one impeller entering the other impeller (figures 3 and 4).

Transfer valve — To make maximum use of the versatility of a multistage pump, a transfer valve is used. This is a two-position valve that allows the impellers to be operated either in parallel (volume) position, or series (pressure) position.

When the transfer valve is placed in the parallel (volume) position, each impeller receives water independently. In effect, it acts as two separate pumps.

In figure 8, the transfer valve has been set for parallel. There is a 20-psi pressure at the inlet to the pump, with each impeller delivering 500 gpm. With the tranfer valve set, the 20-psi pressure forces the flap valve open so that the eye of each impeller has this pressure. Each impeller now builds up the pressure by 150 psi, so that the discharge from each impeller is now 170 psi. The 500 gpm delivered from each impeller reaches the discharge where the total amount of water being supplied by the pump is 1000 gpm at 170 psi.

Now, if the transfer valve were changed to series (pressure), only the first impeller would receive water from the intake (figure 9). The 20-psi intake

Figure 7. Four-stage pump.

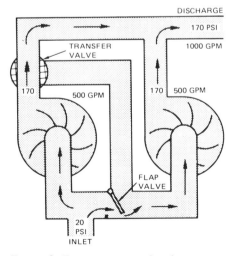

Figure 8. Two-stage centrifugal pump set in parallel.

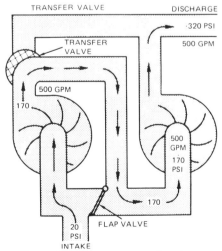

Figure 9. Two-stage centrifugal pump set in series.

pressure at the eye would be increased by 150 psi at impeller 1, so that 500 gpm at 170 psi would flow into the eye of impeller 2. The pressure differential, 170 psi versus 20 psi, would keep the flap valve closed. The 170-psi, 500-gpm flow would be increased at impeller 2 by another 150 psi, so that the discharge would be 320 psi at 500 gpm.

One of the most frequent questions is: When should the transfer be made? Generally, the pump should be operated so that engine rpm is kept within the most optimum range. While for gasoline engines this means at the lowest possible rpm, for diesel engines this might not be the best procedure to follow. Optimum operating rpm for a diesel engine is usually higher than for an equivalent gasoline-driven engine. The pump operator should follow the manufacturer's recommendation whenever possible.

However, as a general guideline, transfer to the parallel position if the pump has to deliver more than 50 percent of its rated capacity. This would be 500 gpm for a 1000-gpm pumper and 625 gpm for a 1250-gpm pumper.

The details of how each manufacturer achieves the series-parallel operation follow:

American LaFrance

The American LaFrance pump is a series-parallel centrifugal type with a volute-shaped discharge. The fire pump is mounted midship on the pumper, behind the engine (figure 10).

The impeller shaft is supported by a floating bearing at the forward end, which is cooled by pump water passing through the clearance spaces and returning to the first-stage suction (figure 11). The rear of the shaft is supported by two ball bearings.

Figure 10. American LaFrance pump mounted on a chassis.

There are two impellers mounted on the shaft between the two sets of bearings (figure 12). On the rear of the shaft are five ring-type shaft packings. Between the two inboard and the three outboard packings is a latern ring that acts as a grease seal (figures 11 and 12).

The hydraulic transfer valve is operated completely by water pressure from the pump, with the position of the valve being controlled by pressure differences in the cylinder and within the header (figures 13 and 14). The valve travels between and seats against valves seats A and B, with corresponding movement of the piston, which is integral with the valve. A selector valve on the pump panel (figure 15) is arranged to admit either pump inlet or discharge pressure into the outer end (area 5 of figure 13) of the transfer valve.

In parallel operation, the selector valve is set at the CAPACITY-PARALLEL position and pump discharge pressure is admitted into the cylinder. Assuming that the transfer valve is at the midpoint of its travel, water at discharge

CENTRIFUGAL PUMPS

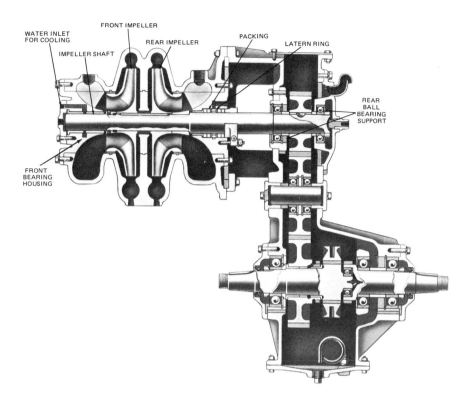

Figure 11. Cross section of American LaFrance pump.

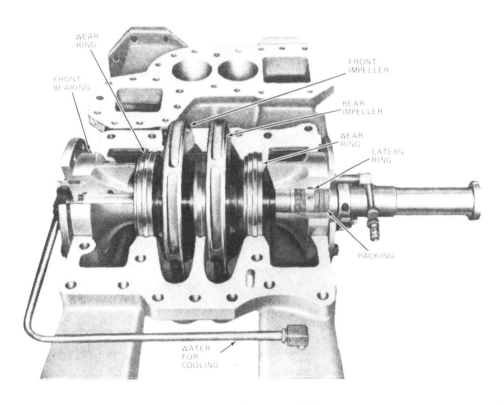

Figure 12. Cutaway of American LaFrance pump.

PUMP OPERATORS HANDBOOK

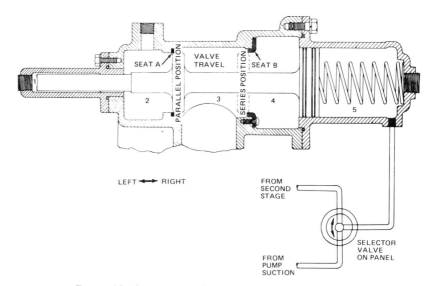

Figure 13. American LaFrance hydraulic transfer valve.

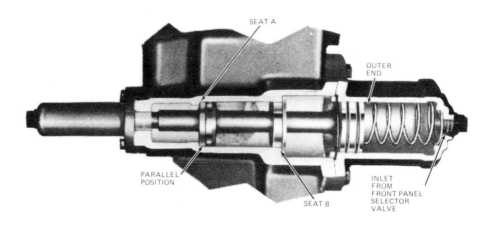

Figure 14A. Cross section of American LaFrance transfer valve.

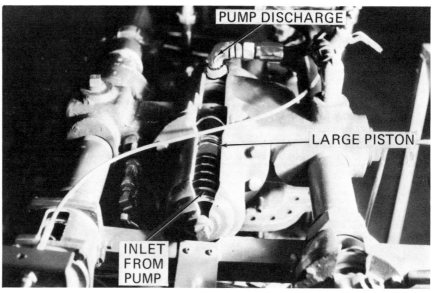

Figure 14B. Cross section of American LaFrance transfer valve.

CENTRIFUGAL PUMPS

Figure 15. Selector valve on pump panel.

pressure is now flowing through areas 1, 2, 3, 4 and 5 of figure 13.

However, the area on the face of the piston at area 5 is much larger than the face of the other piston. With the pressures the same, the force in pounds per unit area will be greater at the larger piston, and the piston will move to the left until the valve seats against A (figure 16). Areas 1 and 2 (figures 13 and 16) will be at inlet pressure, while areas 3, 4 and 5 will be discharge pressure. Due to the difference in area, an even greater force is created to hold the valve in the capacity position.

For series operation, the selector valve is set at the PRESSURE-SERIES position and cylinder 5 (figures 13 and 16) is opened to the pump intake line. Discharge pressure in chambers 3 and 4 will then cause an unbalanced force, due to the difference in area between the valve head and the piston. This unbalanced force starts the valve moving to the right.

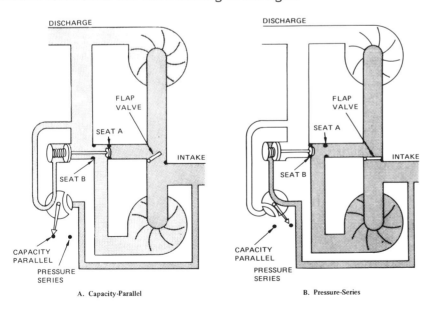

A. Capacity-Parallel   B. Pressure-Series

Figure 16. Schematic diagram of American LaFrance transfer valve.

PUMP OPERATORS HANDBOOK

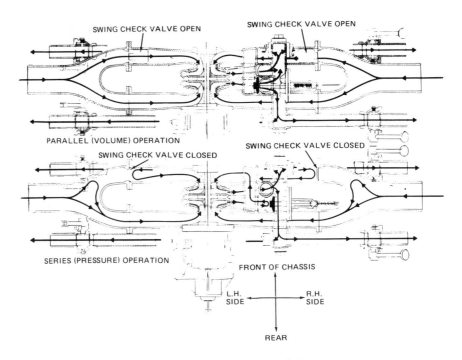

Figure 17. Water flow in the American LaFrance pump.

Now water under discharge pressure (figures 13 and 16) will enter chambers 1 and 2, increasing the force. The valve completes its travel, seating against seat B.

Figure 17 shows that in series position, there are three chambers at different pressures. Chambers 1, 2 and 3 are at first-stage pressure, chamber 4 is at second-stage pressure, and chamber 5 is at intake pressure. Through differences in the area on which they act, these three pressures combine to form an unbalanced force to hold the transfer valve in the series position.

A. Cutaway view of pump

B. Impeller arrangement

Figure 18. American fire pump.

American

The American Pump Company is the only major manufacturer producing a multistage pump which contains two different size impellers, each mounted on a separate shaft (figure 18). The valve is arranged so that the larger impeller

CENTRIFUGAL PUMPS

Figure 19. American transfer selector.

can deliver a large volume, while the smaller impeller produces reduced flow in the area of 200 psi. The two impellers can then be operated in series for pressures of 250 psi or higher.

When the pump selector (figure 19) is set in the series S position, both impellers are engaged. Water enters the first impeller, the pressure is increased, and the water is discharged into the second impeller (figure 20A). The increased pressure forces the flap valve closed. Impeller 2 increases pressure and discharges it into the manifold, thus forcing open poppet valve 2. Poppet valve 1 is kept closed by the difference in discharge and intake pressure.

When the selector is set to the P position, for pressure operation, only the pressure impeller turns (figure 20B). Since impeller 1 is not turning, the intake pressure forces the flap valve open. Impeller 2 increases the pressure, which forces open poppet valve 2 and permits the water to flow into the discharge manifold. Poppet valve 1 is kept closed by the pressure difference between discharge and intake.

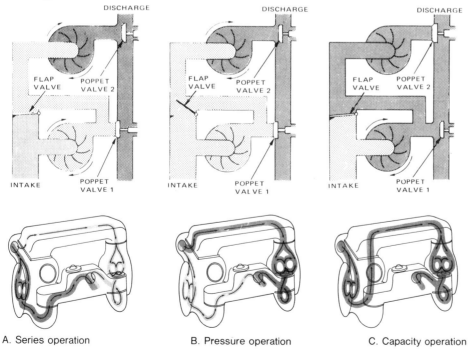

A. Series operation   B. Pressure operation   C. Capacity operation

Figure 20. Operation of the American transfer control.

By setting the selector to C, for capacity operation, only the capacity impeller turns (figure 20C). Intake pressure enters impeller 1, pressure is increased, and water is discharged into the manifold, forcing open poppet valve 1. The discharge from impeller 1 also is used to keep the flap valve closed. Poppet valve 2 is kept closed by the pressure differences between discharge and intake.

The poppet valves (figure 21) are necessary to prevent water under pressure from the discharge from flowing back into the other impeller. These valves also allow the priming of the pump with the discharge valves open.

Figure 21. American pump poppet valve.

Darley

The Darley type S or SH pumps (figure 22) are either two stages or three stages and are mounted midship. Water flow through the pump (figure 23) is directed from the pump panel by a rod which controls the transfer valve. When the rod is out, the valve is in the volume position.

In the parallel position, the transfer valve allows water to flow to both impellers while blocking the crossover (figures 23 and 24). This is actually a brass valve in a brass sleeve.

When switching to series, the output of the first-stage impeller is blocked from going to the discharge and instead is routed to the intake of the second impeller (figure 25). In addition, the increased intake pressure to the second

Figure 22. Darley pump.

CENTRIFUGAL PUMPS

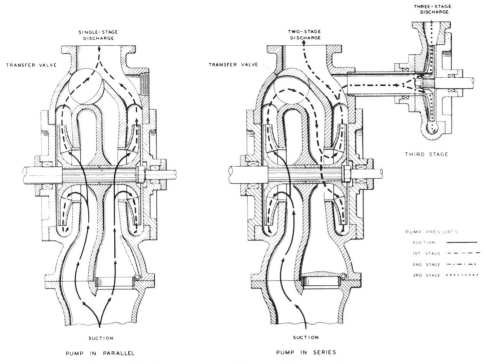

Figure 23. Flow diagram of Darley Champion fire pump.

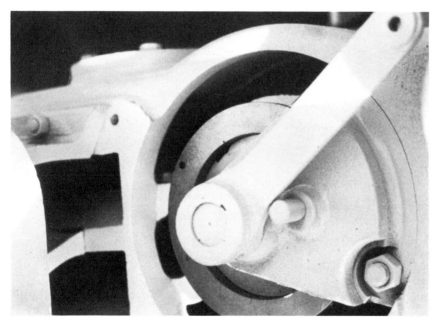

Figure 24. Darley transfer valve.

impeller closes the first-stage check valve. This valve (figures 26 and 27) is forced open during parallel operation.

The Darley transfer valve can also be operated by vacuum or air-powered devices as an option.

Hale

The Hale two-stage pump is designed so that the discharge from each impeller occurs on opposite sides and supplies opposing volutes (figure 28).

To control flow, Hale uses both a manual and power transfer valve. The transfer valve uses the water developed within the pump to perform the

Figure 25. Darley transfer valve.

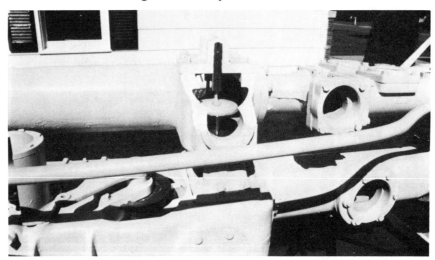
Figure 26. Darley check valve open.

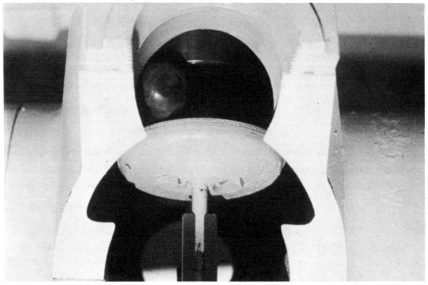
Figure 27. Close up of Darley check valve in the open position.

CENTRIFUGAL PUMPS

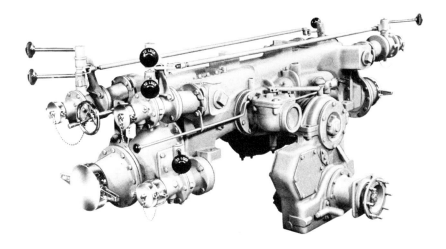

Figure 28. Hale two-stage pump.

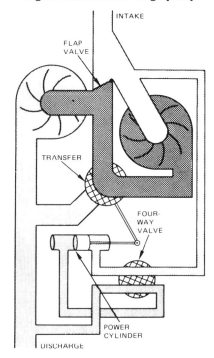

A. Schematic diagram

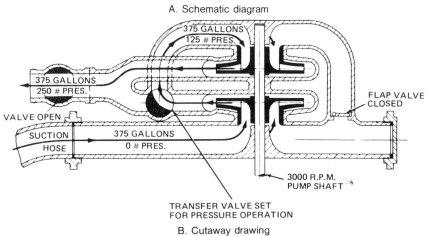

B. Cutaway drawing

Figure 29. Hale pump in pressure position.

transfer. A four-way valve (figures 29A and 29B) has a supply line from the pump discharge and one from the intake. It also has two lines connected to a power cylinder, one to each end of the water chamber. Figures 29A and 29B show the pump operating in the pressure mode, with the discharge pressure keeping the cylinder in position.

To transfer the pump, the four-way valve is switched to the volume position. This reverses the position of the pressure lines to the power cylinder (figures 30A and 30B). With the discharge pressure at the bottom and only intake pressure at the top, the power cylinder moves up and the transfer occurs. The transfer valve will only operate if water is flowing.

If for any reason this valve will not operate, it can be turned manually, either by using a wrench on the hexagonal shaft (figure 31) or by inserting a 3/8-inch rod in the hole provided in the shaft. The shaft can then be turned until the indicator shows the desired position.

Several variations on this power system have been developed by Hale. Some pumps are supplied with power operations that use compressed air from the apparatus air brake system to accomplish the transfer. In these cases, a valve similar to the one in the hydraulic transfer is used, but it is supplied with

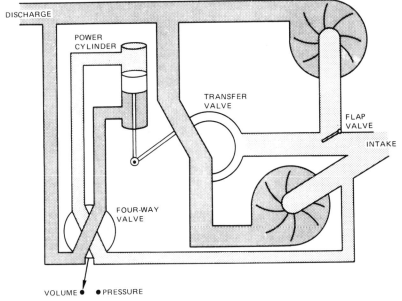

A. Schematic diagram

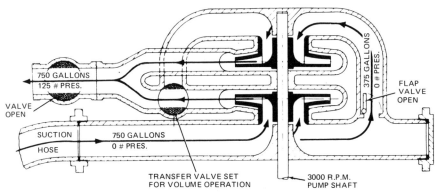

B. Cutaway drawing

Figure 30. Hale pump in volume position.

air pressure on one side and atmospheric pressure on the other. This pressure difference, usually about 90 psi, will cause the piston to move as described above.

In another variation, some Hale pumps use a vacuum-type transfer (figure 32). This system uses the supply of vacuum from the intake manifold of the truck. The pressure or volume button is depressed, and the engine vacuum line is interchanged with atmospheric pressure between the ends of the cylinder, causing the transfer to occur.

Water flow in the series and parallel positions is shown in figures 33 and 34.

The Hale single-stage pump delivers the required water by using a double-suction, single impeller (figure 35). Because it has only one impeller, no transfer valve is required. Flow through the single-stage pump is shown in figure 36.

The single-stage pump also uses the opposed discharge volute cutwaters to ensure radial hydraulic balance (figure 37).

Figure 31. Hale pump panel transfer valve indicator.

Figure 32. Hale pump panel vacuum transfer valve.

Seagrave

The Seagrave fire pump is a two-stage, pressure-volume type, with two impellers mounted on the same shaft. By operating a transfer valve, the pump can be used to supply rated capacity at reduced pressure or to supply reduced capacity at high pressures (figure 38).

The pump shaft is supported by three water-lubricated bearings and a ball bearing at the drive end to take any unbalanced end thrust.

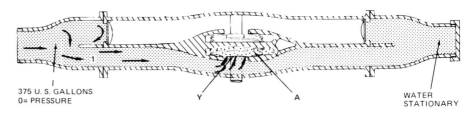

FIRST STEP IN SERIES OPERATION

375 U.S. GALLONS OF WATER ENTER PUMP AT LEFT, FLOWING THROUGH SUCTION CHANNEL #1. THIS WATER IS COLORED GRAY. GRAY WATER IS DRAWN INTO FIRST IMPELLER (MARKED A) THROUGH OPENING "Y", THE EYE OF THE IMPELLER.

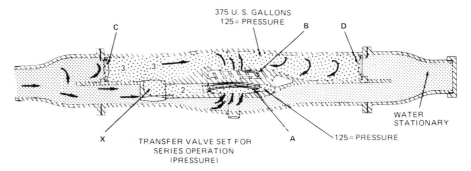

SECOND STEP IN SERIES OPERATION

AS WATER ENTERS FIRST IMPELLER A, WE COLOR IT ▨
▨ WATER LEAVES IMPELLER A AT EDGE OF IMPELLER AND ENTERS UPPER CHANNEL #2 AT 125 LBS. PRESSURE. TRANSFER VALVE X IS SET FOR SERIES OPERATION. WATER PASSES DOWN THROUGH VALVE X AND ENTERS LOWER CHANNEL #3 ON FAR SIDE OF PUMP. CHECK VALVES C AND D ARE CLOSED BY PRESSURE OF WATER. ▨ WATER ENTERS SECOND IMPELLER B, SHOWN IN ▨

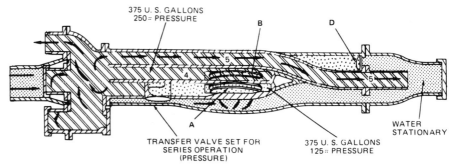

THIRD STEP IN SERIES OPERATION

AS WATERS SECOND IMPELLER B, WE COLOR IT▨
▨ WATER LEAVES IMPELLER B AT TOP EDGE AND ENTERS CHANNEL #4 WITH PRESSURE DOUBLED, I.E., AT 250 LBS.
THE WATER (375 GALLONS AT 250 LBS.) LEAVES PUMP AT EITHER END OF CHANNEL #5, THROUGH DISCHARGE VALVES.

Figure 33. Hale pump operating in series.

The transfer valve is operated hydraulically by pump pressure. A four-way valve connects from the pump discharge and pump intake to a water-operated cylinder which connects to an arm of the transfer valve (figure 38). The two flap valves, one on each side of the pump at the inlet to the second-stage impeller, operate automatically (figure 39).

When the transfer valve is set to the pressure position (figure 40A), the discharge pressure is connected to the cylinder at the right side. This pressure is higher than the intake pressure on the left side, so that the cylinder moves to the left. This connects the discharge of impeller 1 to the intake of impeller 2. The valve which accomplishes this is shown in figure 40B.

To transfer to volume, the valve is switched. This places the high discharge

CENTRIFUGAL PUMPS

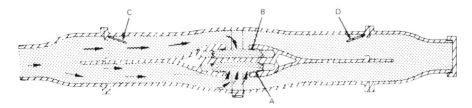

FIRST STEP IN PARALLEL OPERATION
750 U. S. GALLONS OF WATER ENTER CHANNELS #1 AND #3, 375 GALLONS INTO EACH CHANNEL. CHECK VALVES **C** AND **D** OPEN. WATER COLORED GRAY. WATER SHOWS ON ENTERING IMPELLERS **A** AND **B** FROM CHANNELS #1 AND #3 AT SAME TIME.

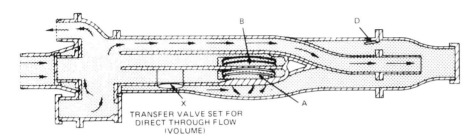

SECOND STEP IN PARALLEL OPERATION
WATER SHOWS ▓▓▓ ON LEAVING EDGES OF IMPELLERS **A** AND **B**. WATER SIMULTANEOUSLY ENTERS UPPER CHANNEL #2 FROM IMPELLER **A** AND UPPER CHANNEL #4 FROM IMPELLER **B**.
TRANSFER VALVE "X" IS SET FOR DIRECT-THROUGH FLOW, FOR PARALLEL OPERATION. A TOTAL OF 750 U. S. GALLONS AT 125 LBS. PRESSURE CONTINUES OUT OF EITHER END OF THE CHANNEL, THROUGH DISCHARGE VALVES.

Figure 34. Hale pump operating in parallel.

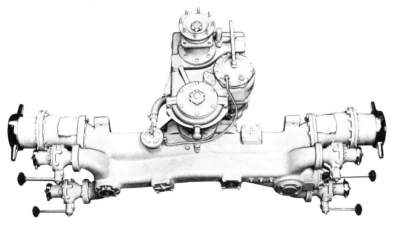

Figure 35. Hale single-stage pump.

pressure on the left side and the low intake pressure on the right side (figure 41A). The cylinder moves to the right and now both impellers are connected to the intake. The valve which accomplishes this is shown in figure 41B.

Water flow from the Seagrave pump is shown in figure 42. For ease of understanding, the diagram only shows inlet on one side and discharge on the other. There are actually inlets and discharges on both sides of the pump.

Waterous

While Waterous manufactures many different kinds of pumps, from single-stage to four-stage, the most popular one for fire service use is the CM-series. This series is a two-stage pump with a transfer valve (figures 43 and 44) de-

109

PUMP OPERATORS HANDBOOK

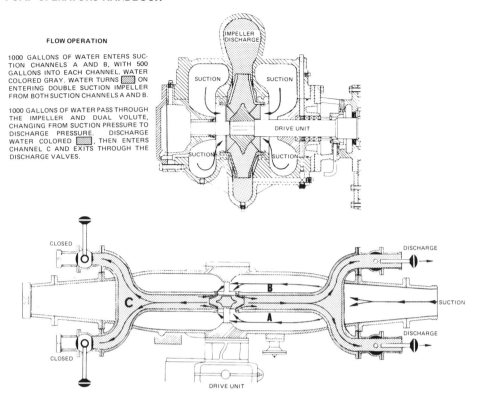

Figure 36. Water flow through the Hale single-stage pump.

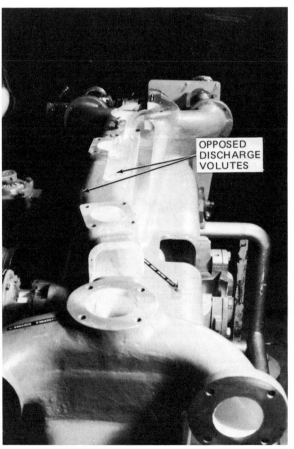

Figure 37. Opposed-discharge volute cutwaters on a Hale single-stage pump.

CENTRIFUGAL PUMPS

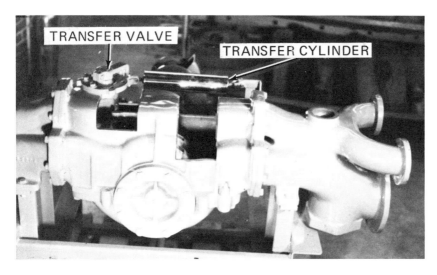

Figure 38. Seagrave fire pump.

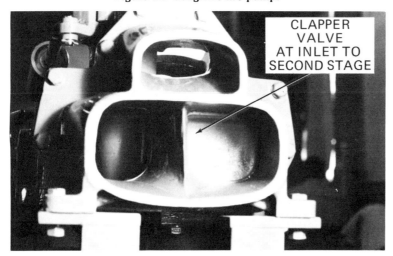

Figure 39. Seagrave flap valve.

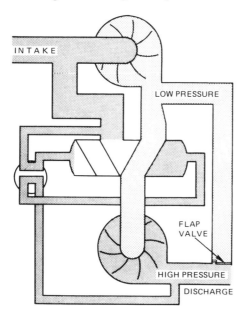

Figure 40A. Schematic diagram of a Seagrave pump operating in pressure.

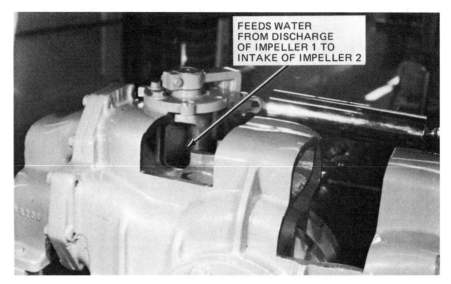

Figure 40B. Water control valve on a Seagrave pump operating in pressure.

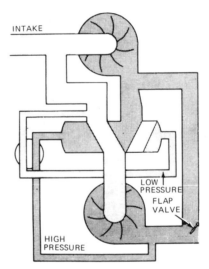

A. Schematic diagram

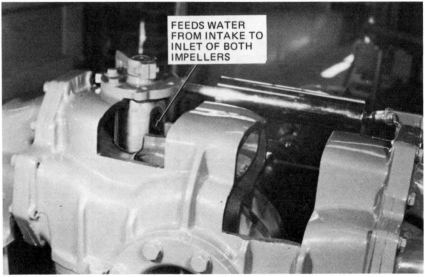

B. Water control valve

Figure 41. Seagrave pump operating in volume.

CENTRIFUGAL PUMPS

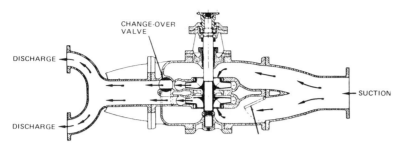

A. Volume position

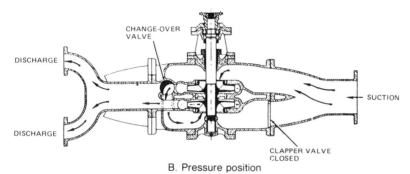

B. Pressure position

Figure 42. Seagrave pump water flow.

Figure 43. Waterous two-stage pump.

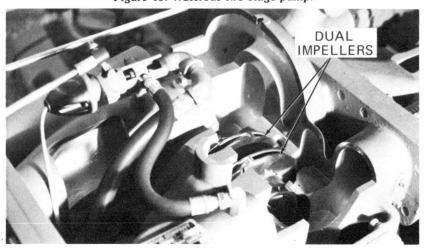

Figure 44. Waterous impellers.

signed for midship mounting. Figures 3 and 4 show the components of the Waterous CM pump.

Transfer is accomplished by an electric transfer valve actuator (figure 45). As an optional feature, a manual control can be connected for operation in case of electrical failure. When the switch is held in pressure position, the cylinder rod extends, rotating the transfer valve (figure 46). This connects the discharge of impeller 1 to the intake of impeller 2 and also forces the flap valve closed (figure 47A). When the valve has completed the transfer, a pressure lamp switch is closed, lighting the pressure lamp (figures 45 and 46).

Figure 45. Waterous transfer valve switch.

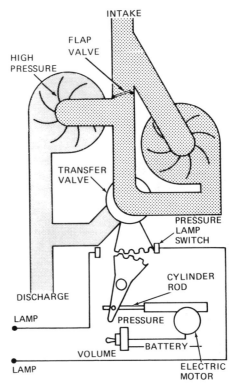

Figure 46. Waterous transfer in pressure position.

CENTRIFUGAL PUMPS

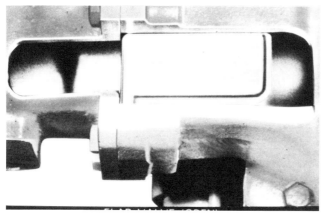
A. Closed

B. Open

Figure 47. Waterous flap valve.

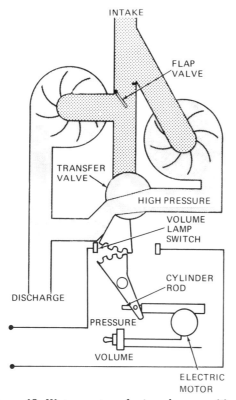

Figure 48. Waterous transfer in volume position.

To transfer to volume, the switch is held down. The cylinder rod retracts, rotating the transfer valve (figure 48). This connects the discharge of both impellers to the manifold of the pump. The flap valve opens due to reduced pressure and the eye of each impeller receives water from the intake (figure 47B). When the valve has completed the transfer, a volume lamp switch is closed, lighting the volume lamp (figures 45 and 48).

Water flow for the volume and pressure positions of the transfer valve is shown in figure 49.

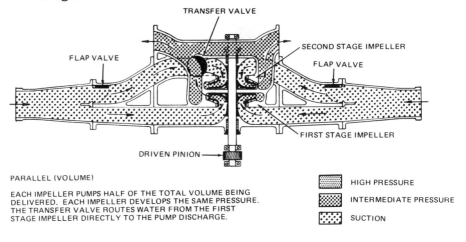

PARALLEL (VOLUME)

EACH IMPELLER PUMPS HALF OF THE TOTAL VOLUME BEING DELIVERED. EACH IMPELLER DEVELOPS THE SAME PRESSURE. THE TRANSFER VALVE ROUTES WATER FROM THE FIRST STAGE IMPELLER DIRECTLY TO THE PUMP DISCHARGE.

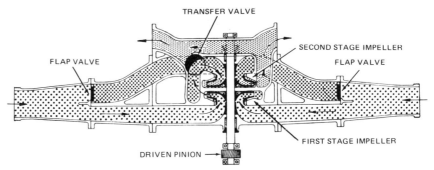

SERIES (PRESSURE)

EACH IMPELLER PUMPS ALL OF THE TOTAL VOLUME BEING DELIVERED. EACH IMPELLER DEVELOPS HALF OF THE TOTAL PUMP PRESSURE. THE TRANSFER VALVE ROUTES WATER FROM THE FIRST STAGE IMPELLER TO THE SECOND STAGE SUCTION. SUCTION FLAP VALVES ARE CLOSED BY THIS PRESSURE.

AT A CONSTANT IMPELLER SHAFT SPEED, WHERE CHANGING FROM PARALLEL TO SERIES OPERATION, PRESSURE IS DOUBLED AND VOLUME HALVED.

Figure 49. Waterous water flow.

Chapter 10

Pump Drives

Now that we've discussed the two types of pumps (positive displacement and centrifugal) and *how* they operate, let's concern ourselves with getting them the power needed *to* operate.

"The fire apparatus driver/operator shall identify three methods of power transfer from the vehicle engine to the pump...

"The fire apparatus driver/operator, given a fire department pumper used by the authority having jurisdiction, shall demonstrate the method(s) of power transfer from vehicle engine to pump."*

Engines used in fire department pumpers are generally those that have been designed for commercial service; although a few manufacturers utilize commercial developments to produce motors exclusively for fire service use. However, no matter what the origin of the motor design, the engine must be chosen to provide the correct torque to drive the apparatus as well as the pump. These engines can have from six to 12 cylinders.

HORSEPOWER

The amount of power or work that can be produced by an engine is measured in terms of the unit horsepower. (While the terms "work" and "power" are not synonymous in scientific terms, they will be used interchangeably in this book.) Horsepower is defined as the amount of work necessary to move 33,000 pounds a distance of 1 foot in 1 minute. This can be shown with the equation:

$$1 \text{ horsepower} = \frac{33,000 \text{ pounds} \times 1 \text{ foot}}{1 \text{ minute}}$$

The common rating for the amount of power which an engine can deliver is brake horsepower (BHP). There are normally two measurements for BHP given for each motor — gross brake horsepower and net brake horsepower.

The *gross brake horsepower* rating, while determined by actual tests, does not accurately reflect the amount of work which the engine can deliver. This is due to the following test adjustments:
1. Temperature at carburetor is corrected to 60°F.
2. Air pressure is corrected to 29.92 inches of mercury.

---

*Paragraphs 3-5.1 and 3-6.1. Reprinted with permission from NFPA 1002-1982, Standard for Fire Apparatus Driver/Operator Professional Qualifications, Copyright©1982, National Fire Protection Association, Quincy, Massachusetts 02269. This reprinted material is not the complete and official position of the NFPA on the referenced subject, which is represented only by the standard in its entirety.

3. Engine is run without fan alternator or generator, air cleaner, governor, or air compressor.

4. Exhaust system altered for more efficient removal of gases.

Each of these changes eliminates items that will decrease the amount of horsepower available for driving the pump. For this reason the *net brake horsepower* more accurately reflects the amount of power that can be produced by a particular engine.

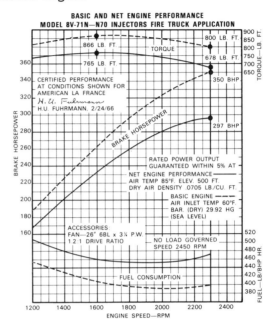

Figure 1. Engine performance curves.

Figure 1 shows an engine performance curve for a Detroit Diesel model 8V-71N engine as modified for fire truck application. The gross brake horsepower for this engine is shown as 350 BHP, while the net brake horsepower is 297 BHP. While the horsepower curves stop abruptly, this does not mean that above 2300 rpm no power is produced. What it does mean is that the engine is governed at 2300 rpm, which is a point slightly below peak power. Above peak power the curve would fall off rapidly, so that an increase in rpm would cause a decrease in brake horsepower.

When only gross BHP ratings are given, an accurate estimate of net performance can be made by subtracting 15 percent of the gross amount from the entire gross amount. This can be expressed as

$$\text{Net BHP} = \text{Gross BHP} - .15 \,(\text{Gross BHP})$$

Example: If the gross brake horsepower is 280 BHP, what is a workable estimate of the net brake horsepower?

Step 1. Determine the correct equation:

$$\text{Net BHP} = \text{Gross BHP} - .15 \,(\text{Gross BHP})$$

Step 2. Determine the formula values:

$$\begin{aligned}\text{Gross BHP} &= 280 \\ .15\,(\text{Gross BHP}) &= 280 \times .15 \\ &= 42\end{aligned}$$

Step 3. Solve the equation:

$$\text{Net BHP} = 280 - 42 = 238 \text{ BHP}$$

One other interesting point is the origin of the term "brake" horsepower. In order to measure the horsepower, a scientific instrument called a *dynamometer* was used to measure force. One of the dynamometers was called the Pony Brake. From the name of this instrument came the term brake horsepower.

TORQUE

A second method of specifying power developed by the engine is called torque. This is the ability of the engine to produce rotation at a given speed. The relationship between torque and horsepower can be expressed as

$$\text{Torque (lb/ft)} = \frac{\text{Horsepower} \times 5252}{\text{rpm}}$$

The torque curves in figure 1 show that 800 pounds/feet of torque is developed for the maximum gross BHP, while 678 pounds/feet of torque is developed for the maximum net BHP. Note that the maximum torque is developed at 1600 rpm.

Example: Using the curves of figure 1, calculate the torque at 1800 rpm, using the net brake horsepower curve.

Step 1. Determine the correct equation:

$$\text{Torque} = \frac{\text{Net BHP} \times 5252}{\text{rpm}}$$

Step 2. Determine the formula values:

$$\text{Net BHP} = 260 \text{ (figure 1)}$$
$$\text{RPM} = 1800$$

Step 3. Solve the equation:

$$\text{Torque} = \frac{260 \text{ BHP} \times 5252}{1800}$$

$$\text{Torque} = 758.6 \text{ lbs/ft}$$

Step 4. Check the results on the curve:

For 1800 rpm, the torque curve shows approximately 760 pounds/feet, using the scale on the right side of the curve.

WATER HORSEPOWER

Just as brake horsepower is an indication of the amount of work available from an engine, water horsepower (WHP) is the amount of work that can be performed by a pump.

The basic horsepower equation states that

$$1 \text{ HP} = \frac{33,000 \text{ pounds} \times 1 \text{ foot}}{1 \text{ minute}}$$

PUMP OPERATORS HANDBOOK

If it takes 33,000 pounds to move 1 foot in 1 minute to equal 1 horsepower, it will take 1/33,000th of the original value to move 1 pound, 1 foot in 1 minute. The general equation for the amount of horsepower developed for each minute of work can then be expressed as

$$HP = \frac{1}{33,000} \times pounds \times feet$$

Every gallon per minute pumped weighs 8.34 pounds, and each psi of pressure developed in equivalent to 2.31 feet. The equation can now be written

$$WHP = \frac{(8.34\ lbs \times gpm)\ (2.31\ ft \times lbs)}{33,000}$$

$$WHP = \frac{gpm \times psi \times 8.34 \times 2.31}{33,000}$$

$$WHP = \frac{gpm \times psi \times 19.27}{33,000}$$

$$WHP = \frac{gpm \times psi}{\frac{33,000}{19.27}}$$

$$WHP = \frac{gpm \times psi}{1713}$$

The efficiency of the pump depends on many factors. It is impossible for all of the horsepower being developed by the engine to produce work by the pump. Losses can be caused by mechanical action and hydraulic friction loss. The mechanical points of contact (drive gearing, shaft bearings, pump packing, and impeller mounting on the shaft) cause a loss of efficiency, while friction loss at the various internal parts of the pump (impeller eye and intake piping) cause additional losses.

The usual efficiency for a fire department pump operating at rated capacity is about 60 to 70 percent. This value is arrived at by calculating the theoretical value and then measuring the actual value being produced. Efficiency can then be determined as

$$Efficiency\ (\%) = \frac{WHP_{out}}{BHP_{in}}$$

To find what brake horsepower (BHP_{in}) is needed to produce a particular water horsepower out (WHP_{out}) for a given pump with a known efficiency, the equation can be rewritten as

$$BHP_{in} = \frac{WHP_{out}}{Efficiency}$$

Example: What brake horsepower is needed to drive a 1000-gpm pump at 150 psi if it is assumed to be 70 percent efficient?

Step 1. Determine the correct equation:

$$BHP_{in} = \frac{WHP_{out}}{Efficiency}$$

$$BHP = \frac{psi \times gpm}{1713 \times efficiency}$$

Step 2. Determine the formula values:

Pressure = 150 psi
Flow = 1000 gpm
Efficiency = 70%

Step 3. Solve the equation:

$$BHP = \frac{150 \times 1000}{1713 \times .7}$$

$$= \frac{150,000}{1199.1}$$

$$= 125.1$$

SINGLE-STAGE VERSUS TWO-STAGE PUMPS

A typical efficiency curve for a single-stage pump is shown in figure 2. Maximum efficiency of 71 percent at 150 psi is obtained with 900 gpm flowing. As previously explained, there are losses in efficiency due to water flow at the eye of the impeller and at the intake piping. Since a single-stage pump with a single-suction impeller has only one impeller eye, there is less loss. It is estimated that at 250 psi about 12 hp is saved. In addition, the larger intake piping required for a single-stage pump means less internal friction loss and the larger volute helps reduce the friction.

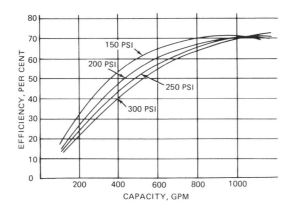

Figure 2. Single-stage pump efficiency curve.

The two-stage pump (figure 3) covers a wider range of pressures more efficiently than does a single-stage pump. Two sets of curves are shown because in reality it is two separate pumps. As shown on the two-stage curves, the average efficiency over the entire operating range is high. Because the general overall operating efficiency is higher for the two-stage pump, less horsepower is needed to cover the range of operation. This becomes especially important during booster line operation where the high pressure and small flow are need-

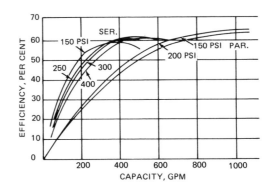

Figure 3. Two-stage pump efficiency curve.
Copyright by "Fire Engineering" Dun-Donnelley Corp, New York, N.Y. March 1962. (Reprinted with permission)

ed. The single-stage pump would need a higher rpm, which means that a greater loss occurs within the pump, thus resulting in the heating of the internal parts of the pump. Since the two-stage pump operates more efficiently at the lower flow and higher pressure, less heating will occur.

PUMP SPEED

Pump speed is determined by holding a counter at the pump speed connection at the pump panel. However, this is not always the most practical method of determining pump speed. The speed can be determined mathematically by multiplying the engine rpm (as read on the tachometer) by the engine-to-pump gear ratio.

The gear ratio selected enables a particular pump to be matched to a particular engine and have the pump rotate so that it will deliver the flow required over the pressure range necessary. Typical gear ratios range from 1.4 to 2.1

Example: Suppose a 1000-gpm pumper is delivering 750 gpm at 150 psi with an engine to pump ratio of 1:1.86 and an engine speed of 1690 rpm. What pump speed will be necessary for the pump to deliver 1000 gpm at 150 psi?

Step 1. Determine the flow to speed proportion:

$$\frac{750 \text{ gpm flow}}{\text{speed at } 750} = \frac{1000 \text{ gpm flow}}{\text{speed at } 1000}$$

Step 2. Determine the formula values:

Flow at 750 = 750 gpm
Speed at 750 = 1690 rpm
Flow at 1000 = 1000 gpm

Step 3. Solve the equation:

$$\frac{750}{1690} = \frac{1000}{\text{speed at } 1000}$$

$$\text{Speed at } 1000 = \frac{1690 \times 1000}{750}$$

$$= 2253 \text{ rpm}$$

Pump speed = engine speed × gear ratio
= 2253 × 1.86
= 4190 rpm

TRANSMITTING POWER TO PUMP

There are four basic ways of supplying power to the pump. Each has advantages and disadvantages, depending on the use of the pump, its location on the truck in relation to the engine, and the pump ratings needed. The four methods are:

Operation from a separate engine,
Operation from the front of the crankshaft,
Operation from the drive shaft,
Operation from a power take-off (PTO) from the engine.

Separate engine — Separate engines are usually used to power small, portable pumps (figure 4). However, they are also found on tankers, small brush vehicles, and airport crash trucks.

Advantages
1. Pump can be carried to a water source that is inaccessible to the truck.
2. Pump speed independent of vehicle speed. Can pump and drive at the same time.

Disadvantages
1. Limited capacity and pressure.
2. Additional engine to maintain.
3. Carry an extra supply of fuel.

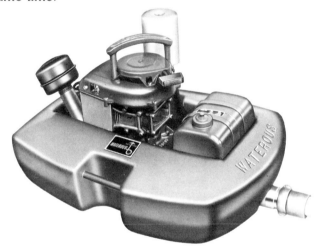

Figure 4. Separate engine powered pump.

Front of the crankshaft — Pumps mounted on the front of a chassis can be driven through a clutch arrangement from the front end of the crankshaft. A typical clutch arrangement is shown in figure 5. To shift the clutch, most pumps use a lever that engages the clutch (figure 6) and provides power to the pump.

Advantages
1. Simple linkage.
2. Simplified controls.
3. All operations in front of truck.
4. Independent of drive system to rear wheels. Can pump while moving.

Disadvantages
1. Pump subject to freezing and damage because it is out front.
2. Size of pump limited.
3. While moving, pump discharge depends on engine speed.
4. Clutch can slip.

Drive shaft operation — The most common type of pump drive is the one for midship-mounted pumps. In this type of drive, the pump transmission is inserted in the drive line between the engine and rear wheels. When

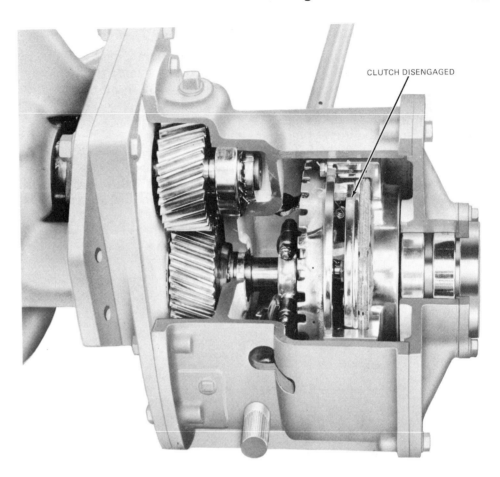

Figure 5. Front-mount clutch arrangements.

Figure 6. Front-mount pump shift.

operating on the road, power is supplied to the rear wheels via the drive shaft. To drive the pump, the power is redirected from the rear wheels and is switched to the pump.

The shifting can be accomplished manually, electrically, or with a vacuum. The method used depends on the individual manufacturer. The transmission must be in direct drive, usually fifth gear, to operate the pump.

One method for transferring the power is through a sliding collar. When the collar is in the road position (figure 7A), the teeth on the coupling shaft transmit the full engine power to the rear wheels. When shifted to pump, the collar moves forward and engages with the pump drive gear, transmitting the engine power to the impeller shaft (figure 7B).

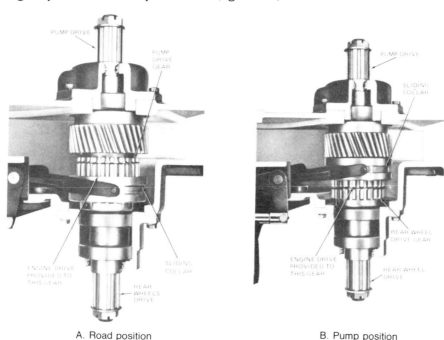

A. Road position   B. Pump position

Figure 7. Sliding collar drive.

One engine manufacturer uses an oil lubricated chain drive to rotate the pump (figure 8A). When in the road position, the engine drives the rear wheels, bypassing the chain drive (figure 8B). When shifted to pump (figure 8C), the chain drive is rotated, thus turning the pump drive. The manufacturer states that there is less loss in the chain arrangement than in the gear drive, so that more usable horsepower is transferred to the pump.

Advantages
1. Full power of engine is available for pumping.
2. Can be used for large size pumps.

Disadvantages
1. Power to drive wheels is disconnected.
2. Relatively complex mechanical operation.
3. Manual override is necessary for electric shift operation.

Another manufacturer has developed a method of powering the rear wheels as well as the midship pump. Mounted on the chassis drive line, the gear case is capable of transmitting the engine torque, as multiplied by the transmission, to the rear axle for road operation. With the transmission in direct drive, the gear case can transmit engine torque to the impeller shaft for stationary pump-

PUMP OPERATORS HANDBOOK

A. Overall view

B. Road position

C. Pump position
Figure 8. Chain-driven drive.

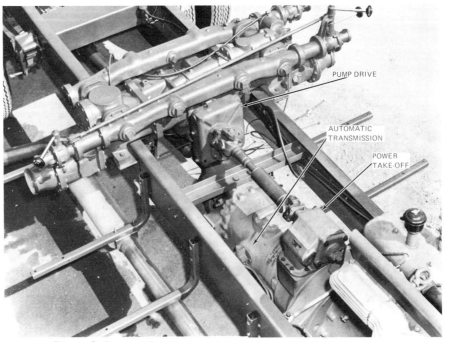

Figure 9. Power takeoff pump drive with automatic transmission.

ing. With the transmission in low (first gear), the gear case can transmit torque to the pump impeller shaft as well as to the rear axle for pump and motor operation.

Power take-off — Another common pump drive for a brush truck is a flywheel power take-off operation. This PTO unit is inserted between the engine and transmission and runs at a constant speed with the engine. This type of PTO is usually capable of handling full engine torque so it is also used to drive full-size pumps. Like the front-mount pump, a clutch is required for engaging and disengaging the pump. For a brush truck, it is desirable that this clutch be operable from both inside the cab and from the pump panel while the engine is running at pumping speeds.

A truck with a midship pump and automatic transmission must be driven by a power take-off from the transmission. The Allison HT-70 transmission (figure 9) has a PTO aperture available with torque capacity to drive a major pump. This PTO is a flywheel type and is not affected by shifting of the transmission. This makes it suitable for pumping while moving.

Advantages
1. Can pump in motion.
2. Simple linkage.
3. Can be used to drive large pumps.

Disadvantages
1. For some engines, a limited amount of power is available.

Chapter 11

Pump, Cab, Body Components

For the operator of a fire department pumper to produce the necessary quantity of water at the desired pressure, he must have a knowledge of all the controls in the apparatus cab and at the pump panel, as well as a working knowledge of the equipment carried on the pumper. The difference between a lever puller who opens the lower left valve one turn every time and the pump operator who opens the auxiliary cooling valve one turn while watching the engine temperature gage can be the difference between success and failure on the fireground.

"The fire apparatus driver/operator, given a fire department pumper, shall identify all gages and demonstrate their usage.

"The fire apparatus driver/operator shall identify the auxiliary cooling systems, and show their function...

"The fire apparatus driver/operator, given a fire department pumper, shall locate, identify, and demonstrate the operation of all equipment carried on or attached to that pumper.

"The apparatus driver/operator shall identify the characteristics and limitations of hard suction and soft suction pumper supply hose (see Chapter 16 also)...

"The fire apparatus driver/operator, given a fire department pumper, shall demonstrate the operation of the auxiliary cooling system."[*]

Correct operation of a pumper requires a complete understanding of how the controls operate and when to use them. Failure in this area can endanger rescue operations and fellow firefighters.

Motor gages

Every time an operator drives the pumper, he should carefully observe all the motor gages in the cab. These gages provide information on the condition and ability of the engine to get the apparatus to the scene and to power the pump. In addition, before leaving the cab at the fireground, the operator should again scan each of the motor gages. Troubles noted at this time can avoid an emergency shutdown situation when firefighters are inside the building.

---

[*]Paragraphs 3-5.4, 3-5.5, 3-6.8, 3-6.9 and 3-6.12. Reprinted with permission from NFPA 1002-1982. Standard for Fire Apparatus Driver/Operator Professional Qualifications, Copyright© 1982, National Fire Protection Association, Quincy, Massachusetts 02269. This reprinted material is not the complete and official position of the NFPA on the referenced subject, which is represented only by the standard in its entirety.

PUMP, CAB, BODY COMPONENTS

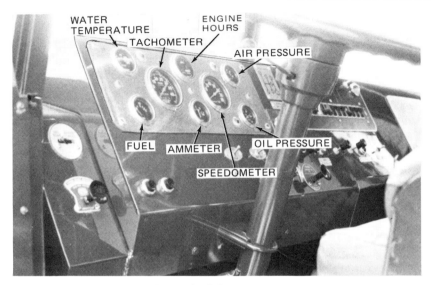

Figure 1. Cab components.

Each apparatus builder has a standard method of locating the cab gages based upon the particular chassis. A typical layout is shown in figure 1. Each operator must become familiar with the gage location for this particular apparatus.

Speedometer — The speedometer indicates the speed of the apparatus in miles per hour. In addition, for pumps that are operated with the shift in gear, the speedometer will read while pumping. Operators should become familiar with the reading for their apparatus so that a quick check can be made to determine if the pump is being operated in the wrong gear.

Contained within the speedometer is the odometer which records the distance traveled in miles. Again, for pumps operated in gear, the odometer continues to function while pumping.

Tachometer — The tachometer indicates the speed of the engine crankshaft in revolutions per minute. The operator should watch the tachometer while driving because it provides an indication of when to shift gears.

When pumping, the tachometer shows how well the pump is being powered:
1. By indicating when the engine is in the wrong gear;
2. By indicating if the transfer valve is in the wrong position;
3. By indicating when there is water slippage from the discharge back to the intake;
4. By indicating when the engine has reached governed rpm;
5. By indicating if there is impairment in the water supply.

Engine hours meter — The engine hours meter records the total time that the engine has been operated. This meter provides an accurate estimate of when preventive maintenance should be performed on the engine.

Fuel gage — The fuel gage records the amount of gas or diesel fuel remaining in the tank. Whenever the gage shows a reading of three-fourths or less, the tank should be filled to ensure that a sufficient amount of fuel will be available if a prolonged pumping operation is necessary.

Oil pressure — The oil pressure gage measures the amount of pressure in the lubricating system. It is important to remember that the gage does *not* indicate the quantity of oil in the lubricating system. As long as there is oil in the system, there will be a pressure reading. Operators should be familiar with the recommended readings for their particular apparatus, so that abnormalities can be readily recognized.

129

If the engine is running at less than 1000 rpm, the gage will normally read less than its recommended value. However, if the gage rises and falls intermittently, this is an indication that the quantity of oil is dropping. If this condition occurs, or if the pressure drops completely, the engine must be stopped immediately (consistent with the hose lines operating in the building) before damage occurs.

Water temperature — The water temperature gage indicates the temperature (in degrees Fahrenheit) of the water in the engine cooling system. This gage is a metallic thermometer element inserted permanently in the cooling space surrounding the engine cylinders. It is important that the engine operate at the correct temperature (about 160 to 190°F) for maximum efficiency.

Too low a temperature can be caused by:
1. Too large a cooling system;
2. Thermostat or shutters remaining open;
3. Circulator or auxiliary cooler valve open too wide;
4. Radiator fill valve open.

Too high a temperature can be caused by:
1. Water level in the radiator too low;
2. Leak in the cooling system;
3. Thermostats or shutters not opening;
4. Auxiliary cooling system not functioning.

A description of the auxiliary cooling system and the associated valves is covered later in this chapter.

Air pressure — Apparatus equipped with air brakes contain a gage that shows the pressure available in the brake lines. When the pressure falls below a certain minimum amount, usually 90 psi, the compressor automatically activates to refill the tank. Should the pressure continue to drop below 60 psi, most brakes will automatically lock and an audible warning device will be sounded, indicating a dangerous situation. Some apparatus have a reserve air tank, that can be manually activated to increase the pressure and release the brakes.

The air horns on apparatus should only be operated by the driver so that he can watch his air pressure gage and discontinue blowing the horn if the pressure should drop.

Ammeter — A properly operating ammeter can indicate how the electrical system is operating and, to some extent, the system's condition. The ammeter indicates how much current is flowing into (charge) or out of (discharge) the battery. In most cases, all electrical equipment except the starter, which draws too much current for the ammeter to handle, is connected through the ammeter.

After a battery has taken as much charge as it can, the voltage regulator cuts off the generator/alternator output, turning it on briefly to peak the battery at frequent intervals. This results in a small pulsation of the ammeter pointer from zero to charge and back again.

The experienced operator recognizes that charge readings can mean:
1. The generator/alternator is charging the battery and supplying enough current to run the electrical equipment as well; or
2. The regulator is not operating properly, which can damage the battery, the generator/alternator, and the electrical equipment in use.

Discharge readings can mean:
1. Insufficient generator/alternator speed due to low motor speed or slipping belt;
2. Generator/alternator inoperative due to burnout, broken drive belt, bad brushes, etc.;

PUMP, CAB, BODY COMPONENTS

3. Low generator/alternator output due to loose connection, bad brushes, open winding, etc.;

4. Inoperative or improperly operating voltage regulator; or

5. Electrical overload, beyond capacity of generator/alternator to supply, due to short circuit or possible mechanical overload, or more equipment than the system was designed to operate.

As a vital part of the total operating system, the electrical parts should be checked frequently to ensure reliability and safety.

Specific gravity readings of a full charged battery should be within .02 of the manufacturer's specification; a difference of .025 between cell readings calls for replacement of the battery.

Pump gages

Two gages necessary for proper pump operation are provided with all fire pumps. One is a *compound gage* for the intake side of the pump, either positive or negative (vacuum) pressure, in the pump intake chamber (figure 2). The second gage is a *pressure gage* that registers pump discharge pressure. For a centrifugal pump, the second gage may be compound type, or it may be a pressure type without the stop at the pressure position (figures 2 and 3). Omis-

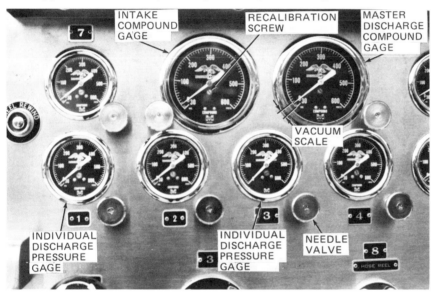

Figure 2. Pump gages.

Figure 3. Master discharge pressure gage.

sion of the stop pin is necessary when using the pressure gage due to the negative pressure (vacuum) throughout the pump during the priming operation. (The centrifugal pump is a continuous waterway.)

Both gages are activated by a hollow curved tube known as a *Bourdon tube* (figure 4). When a vacuum is being created in the pump, as during a priming operation, the curve of the Bourdon tube decreases, and this movement is transmitted through the linkage to register the vacuum. The vacuum reading is shown on the gage in inches of mercury from 0 to 30 (figure 2).

When a positive pressure (above atmospheric) exists in the pump, the Bourdon tube tends to straighten. The movement will then be proportional to the positive pressure and will be registered by the indicator on the dial, as pounds per square inch pressure. Remember that a 0 psi reading on the gage is really 14.7 psi absolute pressure at sea level.

This gage is standard for fire service use. It is reasonably rugged, but it can be thrown out of calibration and damaged by improper control of the fire pump and by freezing. Fast shutoff of nozzles while the pump is discharging induces shock loading in the pump, causing water hammer which affects the pressure gages. This often happens when operating from water mains where the residual or inlet pressure is in the 10 to 20-psi range.

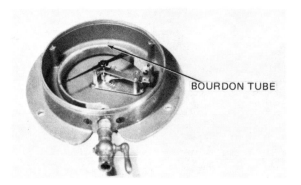

Figure 4. Internal construction of a gage.

Freezing of the gage usually results in permanent deformation of the Bourdon tube. The tube curvature is reduced, resulting in a higher pressure being indicated, and the needle does not return to the zero setting when the pump is shut down. A slight deformation can be tolerated, provided that the excess reading is not greater than 5 psi. This error in reading will not be constant over the operating range of the gage, and any error greater than 5 psi should be corrected by recalibration (figure 2). Some gages can only be recalibrated by resetting the needle and using a gage tester to check. A new gage should read within ± 3 psi at all pressures.

Some gages are the nonrepairable type, and damage to the Bourdon tube requires gage replacement.

Because of the problems associated with the freezing of pumper gages, a new type of gage has been developed (figure 5). In this type, the face of the gage is completely filled with a low-temperature liquid, usually glycerine or silicone.

The advantages of filling the face of the pressure gage with a liquid include:
Reducing freezing problems;
Reduction of needle vibration which makes gages difficult to read;
Reducing corrosion and vibration wear on the gage parts, as well as lubricating the internal parts; and
Reduction of condensation problem on the inside of the gage bezel because the gage is completely sealed.

PUMP, CAB, BODY COMPONENTS

Figure 5. Liquid filled pressure gage.

Another option available with this type of gage is to have it filled with oil and then completely sealed. The water then acts on the oil to produce a pressure reading on the gage. This option provides the additional advantage of eliminating the freezing problem for the gage's working parts and eliminating corrosion and blockage problems by keeping dirty water out of working parts.

Flow meters — The major objective of the pump operator is to provide the required flow of water for the particular nozzle and hose layout being utilized. To produce the correct flow, calculations must be accomplished to produce the necessary engine pressure. The techniques for making these calculations are included in Chapter 6.

Now, however, a new device has been developed that directly reads flow. Since the pump does not create water, the flow out of a discharge is the same as the flow out of the nozzle. Therefore, if a flow meter replaced the standard pressure gage, the pump operator would not need to do any calculations (figure 6).

Remember, as explained in Chapter 6, a discharge pressure reading of 150 psi does not indicate what the flow will be at the nozzle end. The actual flow depends on the specific size and type of the hose, its length, the nozzle size, and the elevation difference between the nozzle and the pump. The flow meter can eliminate all those manual calculations.

Differential flow meter — The differential-type flow meter operates by

Figure 6. Auto Scan digital flow meter (right) aids operator in obtaining correct flow for fire stream operations. Also foam system is additionally monitored for correct flow.

133

placing two tubes with small holes in the flow line. One of the tubes faces into the direction of the flow and the other faces with the direction of the flow. The tubes measure the pressure in each tube, calculate the difference between them, and then convert the differential into a flow reading.

Turbine sensor flow meter — The turbine-type flow meter places a turbine propeller into the middle of the flow. The movement of the turbine is translated by mechanical means to a flow reading and thus a gage reading. The problem with this technique is that the turbine inserted in the stream causes a pressure drop, and damage to the turbine can result if it is struck by debris. In addition, large flows, as experienced in the fire service, are not accurately measured.

Paddle wheel flow meter — In the paddle wheel flow meter, the wheel is inserted on the edge of the flow. The rotation of the paddle wheel is converted into electrical signals that are then transmitted to a gage. Also, instead of the regular needle-type gage, digital readouts can be provided. Because the paddle wheel is out of the center of the flow, the possibility of clogging is lessened, and there is less danger of debris in the line striking the paddle wheel.

Intake gage at draft — The intake gage, when operating from draft, will indicate the amount of vacuum in inches of mercury. The exact reading will depend on the amount of lift from the static source, with approximately 1.1 inches of mercury equaling 1 foot of lift when no water is being discharged. As water is discharged, the reading will increase due to friction loss in the hard sleeves.

When water is flowing, a high vacuum reading usually indicates a blockage of the intake strainer. A low vacuum reading usually indicates an air leak somewhere in the system.

Intake gage at positive pressure — When operating from a water source supplied under pressure, and with no water flowing, the intake gage will read positive static pressure in pounds per square inch. As discharge lines are opened and water begins flowing, there will be a drop in the reading caused by the friction loss due to water flow. The resultant pressure, called residual pressure, provides the pump operator with an indication of how many more lines can be supplied from the source. As additional lines are supplied, there will be a further drop in the gage reading. The operator must maintain a reading of at least 5 psi on the intake gage at all times.

Discharge gages — The discharge gage is a positive pressure gage that records the pressure being pumped to individual hose lines. This type of gage is sometimes known as a *line gage*. There should be a gage for every discharge on the apparatus. The amount of discharge pressure is determined by the amount of water flowing, the size of the hose, nozzle in use, and the length of the line. A drop in the reading of the line gage can be caused by:
1. Reduction of water,
2. Opening another nozzle,
3. Burst section of hose, or
4. Changing nozzles.

A sharp increase in readings of this gage can be caused by:
1. A shutdown of a hose line,
2. A reduction in nozzle size, or
3. A change in pattern in certain fog nozzles.

Needle valve — The needle valve for each gage should be closed to a point where the gage gives a steady reading without vibration (figure 2). If a gage needle does not move, check that the valve is not closed all the way.

Auxiliary cooling

As explained earlier, the operating temperature of the engine is extremely important. An engine operating too hot will be damaged, while one operating too cold will cause sludge deposits within the engine.

The normal apparatus cooling system may be inadequate to keep the engine from overheating, especially while operating hard on the fireground with no ram air passing through. For this reason, an auxiliary cooling system is added to the pumper. This system (figure 7) acts as a heat exchanger. Cool water from the pump circulates through a coil, with the water from the radiator on the outside of the coil. Heat is transferred from the engine cooling water to the pump water.

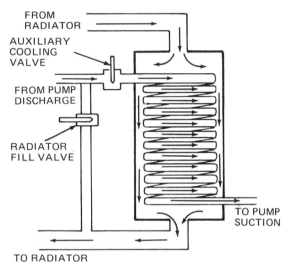

Figure 7. Typical auxiliary cooling valves.

Water from the pump enters the coil by opening the auxiliary cooling valve (figures 7 and 8). Note that the pump water does not mix with the radiator water when the auxiliary cooling valve is opened. Open the auxiliary cooling valve slowly so that the optimum operating temperature can be obtained without cooling the engine too much.

The radiator fill valve admits pump water directly to the radiator (figure 8). This is an emergency device only and should be used with care. Since the incoming water is from the pump discharge under pressure, opening the valve wide will allow a large flow of water to enter the radiator (figure 7).

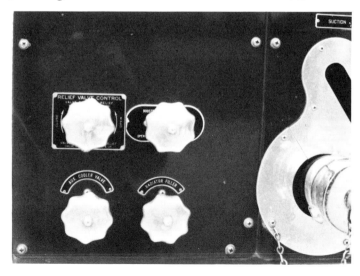

Figure 8. Typical auxiliary cooling valves.

The overflow piping of the radiator may not be sufficient to handle the incoming water from the pump. If the pressure in the radiator builds up, bursting can occur. It is necessary, therefore, to remove the radiator cap before opening the valve. Exercise extreme caution when removing the cap because the original problem was the excess temperature, and as the cap is removed, steam under pressure could be released. Before removing the radiator cap, open the radiator fill valve a small amount, or open the radiator fill valve until water comes out of the overflow piping. Then, the valve is shut and the temperature is checked, eliminating the need for removing the radiator cap.

Discharge and intake valves

Discharge and intake valves regulate the water entering and leaving the pump. They contain a locking device so that if reduced flow from an individual discharge is desired, the valve can be locked in position.

One type of valve is designed for push-pull operation. It consists of a sliding gear-tooth rack that engages a sector gear connected to the valve stem (figure 9). By twisting the T-handle 90 degrees, it is locked in position.

Another type of valve is the quarter-turn valve. As the valve is rotated, the ball rotates from being in-line with the waterway to being 90 degrees to the waterway (figure 10). This valve may be locked in position by twisting the knob on the handle.

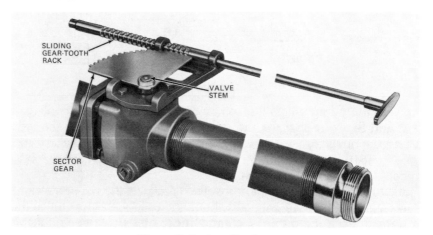

Figure 9. Push-pull valve.

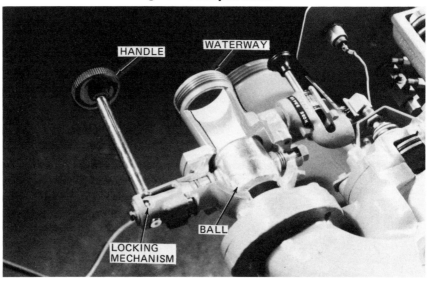

Figure 10. Quarter-turn valve.

PUMP, CAB, BODY COMPONENTS

Drains

Each intake and discharge line should be equipped with a drain. Opening the drain before uncoupling will relieve the pressure in the line and make the task easier. In addition, where freezing is a problem, water must be drained from all the piping.

One simple way to facilitate draining is to use a master drain. Figure 11 shows multiple 1/4-inch ports that are opened by pulling on one plunger.

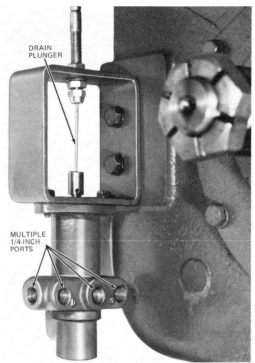

Figure 11. Master drain arrangement.

Water tank

Almost all fire department pumpers currently being manufactured have a water or booster tank. To get water from the tank to the pump, the tank valve must be opened. To allow water to flow from the discharge piping to the pump, the tank fill valve must be opened (figure 12). In addition, most

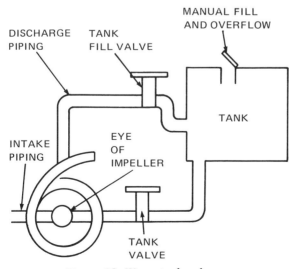

Figure 12. Water tank valves.

137

pumpers have an access port directly to the tank for filling pumpers without having to go through the pump.

When using the tank fill valve to refill the water tank, care must be exercised to be sure that the overflow piping and vent are sufficient to handle the discharge flow and pressure from the pump. If it is not, then the water tank can be deformed and even cracked from the excess water and air pressure.

Pump overheat

The centrifugal pump tends to overheat when turning rapidly and discharging little or no water. To warn the operator that the pump is beginning to heat up, red flashing lights are used at the pump panel (figure 13).

One way to reduce this heat is to open the tank valve and the tank fill valve. This will allow water to pass from the tank through the eye of the impeller, to the discharge piping, through the tank fill valve, back to the tank. This circulation will permit removal of the excess pump heat.

Figure 13. Pump overheat indicator.

Pumper equipment

Pumper equipment varies, based upon the particular needs of individual fire departments. A rural fire department, without a hydrant system, would carry additional hard sleeves for drafting, strainers, drafting basin, as well as additional hose. The fire department in a major city might carry additional preconnect hand lines as well as forcible entry tools. So, while it is difficult to predict exactly what each pumper will carry, much of the equipment is fairly standardized. This section, therefore, covers the common items carried.

Obviously the main objective of the pumper is to move water from the source to the incident scene. To accomplish this objective, the pumper carries hose, nozzles, adapters, control devices, special hose tools.

Hose — Hose varies in size, construction and material. The larger the diameter of the hose, the less friction loss that is created for the same flow. Booster line hose is available in 3/4 or 1-inch diameter and is made of hard rubber. This allows the hose to be rolled onto a reel and charged with water without removing the entire length from the reel. However, flow from booster hose is extremely limited.

Hand lines come in 1½, 1¾, 2 and 2½-inch diameters. These lines provide flows that can be handled with either one, two or three firefighters. The hose is made of fabric such as cotton or man-made fibers or plastic. The covers can be either one or two jackets so that potential use and the ability to withstand abrasions is determined by the type of hose.

In order to move large quantities of water, large-diameter hose is necessary. This includes hose that is 2½, 3, 3½, 4, 5 and 6-inches in diameter. Small lengths of 4, 4¼, 5 and 6-inch hose are used to connect a pumper to a pressure water source. Longer lengths of this hose are used to connect tandem pumpers, a pumper and the supply source, as well as a pumper to the incident. These lengths of short hose ensure an ample flow from a pressure source.

Short 10-foot lengths of 4, 4½, 5 or 6-inch hard sleeve hose are used for drafting water. The stiffening rings of these hoses prevent the hose from collapsing when the inside is under negative pressure during a drafting operation (see Chapter 15, Drafting Operations). Hard sleeves should not be used when operating from a pressure source because the pump operator will not get an indication that the pump is about to cavitate (see Chapter 15).

Nozzles — A pumper carries a wide selection of nozzles to meet various needs. These nozzles include:

Solid stream — hand and master stream
Fog nozzles — single flow, constant flow, variable flow and automatic flow

Solid stream nozzles have a smooth bore opening that produces a solid stream with a great reach between nozzle and point of application. Since the diameter of the nozzle influences the flow (the larger the opening the greater the flow), the nozzle's reaction for holding the line becomes a factor (see Chapter 12, Nozzle Reaction). Therefore, hand lines are considered to be solid stream if they have a tip diameter up to 1¼ inches at 50-psi nozzle pressure. Master stream tips are 1¼ inches and larger and operate at 80-psi nozzle pressure.

Single flow fog nozzles are designed to have their most efficient operation at a specific flow with a 100-psi nozzle pressure. This type of nozzle presents limitations when more water is needed because an efficient stream cannot be developed.

A *constant flow nozzle* produces the specific rated flow, at 100-psi nozzle pressure, regardless of the fog setting selected. This means that as the fog pattern is changed from straight stream to 30° to 60°, the flow remains the same. This results in the same nozzle reaction, and changing patterns do not throw the hose lines personnel off balance.

A *variable flow nozzle* has an adjustment for changing the flow rating. However, just changing the setting does not mean that more water will be produced. What it does mean is that at the new setting *and* with a nozzle pressure of 100 psi, the rated flow will be produced efficiently. The key is that the pump operator must know of the change at the nozzle and adjust the discharge pressure accordingly.

One of the newest designs on the market for nozzles is known as the *automatic flow nozzle*. In this case, the nozzle automatically adjusts its opening to maintain an optimum flow, with a 100-psi nozzle pressure to produce the most efficient stream. In this way, as flow varies, the nozzle continues to provide a good stream. Pump operators can therefore begin water movement quickly, while using a standard discharge pressure, knowing that the created flow will produce an excellent fire stream. The nozzle is designed for both hand lines and master streams.

The easiest way to determine flow with an automatic nozzle is to use flow meters. However, if they are not readily available, the calculations can be done as follows:
Step 1. Determine pump pressure.
Step 2. Subtract the known nozzle pressure of 100 psi from the pump pressure.

Step 3. The result is the pressure available to cover friction loss.
Step 4. Divide the pressure by the number of hundreds of feet of hose.
Step 5. Determine the flow for the friction loss for the hose being used.

Example: A pumper is delivering water to 300 feet of 2½-inch hose with an automatic nozzle. The pump discharge pressure is 208 psi. How much water is flowing?

Step 1. Determine the pump pressure: 208 psi.
Step 2. Subtract known nozzle pressure of 100 psi:

$$208 - 100$$

Step 3. Resultant friction loss is: 108 pounds.
Step 4. Divide the resultant pressure by the number of hundreds of feet of hose:

$$108 \div 3 = 36$$

Step 5. Determine flow for 2½-inch hose with a friction loss of 36 psi per 100 feet: 400 gpm.

Adapters — Hooking up hose lines may, at times, require adapters. These include changing the male end of a hose to a female using a double female adapter. Other adapters may be necessary for mutual aid if the neighboring department uses different threads.

Control devices — For controlling flow in the hose line, several devices are carried on a pumper. The siamese devices carried on a pumper take two or more lines and combine them into a single line. The siamese may have control gates to shut down flow in the lines without shutting down the pumper.

Another control device is a wye which takes a single line and breaks it down into two or more lines. The wye can be gated so flow to the divided lines can be controlled.

Individual gates can also be carried on the pumper. These can be connected to the hydrant for flow control as well as to the pumper for intake control. The gates vary in size from 2½ to 6-inch and operate by quarter turn, gates, or diaphragms.

Special hose tools — There are special hose tools for temporary repairs to broken hose by jacketing it. The jacket clamps around the hose opening and makes a seal at both ends, thus allowing the flow to continue.

A rope hose tool can be used to secure the hose to a ladder and to help hold the hose during times of high flow.

A hose roller can be used to bring the hose over a rough edge without causing abrasion. It also aids in lifting the hose onto a roof.

Entry, ventilation and salvage equipment — Depending upon the needs of the local fire department, the pumper may also carry entry, ventilation, and salvage equipment. This would include:

Entry — axe, pry bar, lights
Ventilation — ladders, pike pole
Salvage — smoke ejector, salvage covers, roof coverings

How each of these tools is used is covered in basic firefighting manuals. The pump operator must be familiar with the portable generator capabilities and ensure that the current load from the lights and smoke ejectors does not exceed the rated amount. In addition, if a separate fuel supply for the generator is necessary, the type and mixture must be known by the pump operator.

Chapter 12

Nozzle Reaction

One of the basic laws of physics, Newton's Third Law, states that for every action there is an equal and opposite reaction. For the firefighter, this means that water flowing out of the nozzle will cause a backward reaction.

Since the reaction force is dependent upon the amount of water flowing through the hose, it will therefore depend on the size of the nozzle used and the nozzle pressure.

"The fire apparatus driver/operator, given a series of fireground situations involving various operating pressures, shall demonstrate the formula for calculation of nozzle reaction of hand and master streams used by the authority having jurisdiction."*

Straight-tipped ground nozzles

The formula for nozzle reaction for straight-tipped solid stream nozzles is:

$$NR = 1.57 \times d^2 \times P, \text{ where}$$
$$NR = \text{nozzle reaction in pounds}$$
$$d = \text{nozzle diameter in inches}$$
$$P = \text{nozzle pressure in psi at the tip of the nozzle}$$

Example: What is the nozzle reaction resulting from a 1¼-inch tip at 50 psi nozzle pressure?

Step 1. Select the correct equation:

$$NR = 1.57 \times d^2 \times P$$

Step 2. Determine the formula values:

$$d = 1¼ \text{ inches}$$
$$P = 50 \text{ psi}$$

Step 3. Solve the equation:

$$NR = (1.57)(1¼)^2(50)$$
$$= 122.7 \text{ lbs.}$$

*Paragraph 3-4.10. Reprinted with permission from NFPA 1002-1982, Standard for Fire Apparatus Driver/Operator Professional Qualifications, Copyright©1982, National Fire Protection Association, Quincy, Massachusetts 02269. This reprinted material is not the complete and official position of the NFPA on the referenced subject, which is represented only by the standard in its entirety.

Actual tests with varous size nozzles show that a 95-pound nozzle reaction is about the maximum that a three-man crew can handle for any length of time.

Straight-tipped ladder pipes

The formula for nozzle reaction of a straight-tipped ladder pipe is the same as for a ground nozzle. However, the pump operator must now be careful to maintain correct pressures to avoid stress on the aerial ladder. The maximum recommended nozzle reaction for a ladder pipe, mounted on the top fly of an aerial ladder, is 400 pounds.

In addition, the flow from a ladder pipe must always be perpendicular to the ladder rungs. If a change in the horizontal direction of the stream is necessary, the ladder should be rotated to avoid placing lateral stress on the ladder.

Example: For a ladder pipe placed on the top fly, can a 2-inch tip with 80-psi nozzle pressure be used?

Step 1. Select the correct equation:

$$NR = 1.57 \times d^2 \times P$$

Step 2. Determine the equation values:

$$d = 2 \text{ inches}$$
$$P = 80 \text{ psi}$$

Step 3. Solve the equation:

$$NR = (1.57)(2)^2 \times (80)$$
$$= 1.57 \times 4 \times 80$$
$$= 502.4 \text{ lbs.}$$

No, this tip should not be used at 80 psi nozzle pressure because the nozzle reaction exceeds 400 psi.

Fog nozzles

The nozzle reaction for variable-pattern nozzles cannot be based on the standard formula because the nozzle diameter does not flow a concentrated core of water. The engineering department of the Elkhart Brass Company has developed a formula for calculating the nozzle reaction of this type nozzle. This formula is:

$$NR = .0505 \times Q \times \sqrt{P}, \text{ where}$$
$$NR = \text{nozzle reaction in pounds}$$
$$Q = \text{flow in gallons per minute}$$
$$P = \text{nozzle pressure in psi, at the base of the nozzle}$$

Example: What is the nozzle reaction for a 1½-inch fog nozzle operating at 100-psi nozzle pressure, flowing 90 gallons per minute?

Step 1. Select the correct equation:

$$NR = .0505 \times Q \times \sqrt{P}$$

Step 2. Determine the equation values:

$$Q = 90 \text{ gpm}$$
$$P = 100 \text{ psi}$$

Step 3. Solve the equation:

$$\begin{aligned} NR &= .0505 \times 90 \times \sqrt{100} \\ &= .0505 \times 900 \\ &= 45.45 \text{ lbs.} \end{aligned}$$

Since many fog nozzles have different flows at 100-psi nozzle pressure, depending on the fog pattern selected, nozzle reaction will vary as the pattern is changed. The reaction and flow will be greatest at a 30° fog pattern.

Water hammer

One other problem with water movement is the danger of sudden stops. Since there is about 25 gallons of water in 100 feet of 2½-inch hose flowing 250 gpm and moving at a velocity of about 14 mph, the sudden stopping of this volume of water causes severe shock loading in the hose, couplings, and pump. These shock loadings are known as water hammer.

For this reason, nozzles, gates, and valves should always be closed slowly. Otherwise, the hose may rupture or the pump may be damaged, causing injury, loss of water, and additional property damage.

Water hammer can also damage the water supply system if a hydrant gate is closed too rapidly. While the hose line can expand and absorb some of the increased pressure due to water hammer, nonflexible, metal pipe, especially that 6 inches or under, has no such elasticity. Broken water mains are a definite possibility of shutting the hydrant down too quickly.

Chapter 13

Pressure Control Systems

To avoid unwanted nozzle reaction and water hammer to protect firefighters from dangerous pressure rises and still provide sufficient hose streams for firefighting operations, the pump operator must be knowledgeable in pressure control.

The term *pressure control* summarizes the responsibilities of the pump operator on the fireground. The operator must first ensure that optimum pressure for the set of conditions existing on the fireground is obtained. He must be sure that the pressure is not too low to deliver sufficient water to the fireground, nor too high to make the hose impossible to handle. Second, he must make certain that this optimum pressure is maintained in each individual line, even though nozzles may be opened and closed.

"The fire apparatus driver/operator shall identify the theory and principles of pumper pressure relief systems and pressure control governors.

"The fire apparatus driver/operator, given a fire department pumper, shall demonstrate the operation of the pumper pressure relief system, or the pressure control governor, or both."*

The method used to regulate the pressure is based on the ways this pressure is developed within the centrifugal pump. These are:

1. Speed of rotation of the impeller;
2. Volume of water moving through the pump;
3. Pressure being supplied to the intake of the pump.

Since the volume of water moving through the pump directly affects the pressure, shutting down a nozzle will cause a sudden increase in the discharge pressure. Even an alert operator will find it impossible to compensate for these surges in time to protect the nozzleman on another line. For this reason, automatic devices have been developed to compensate for these pressure changes.

RELIEF VALVES

Relief valves are used on fire apparatus to prevent excess pressures caused by changes in fireground situations. A relief valve bypasses excess water from the discharge side of the pump back to the intake, thus preventing dangerous pressure surges by changing the volume of water flowing through the pump.

---

*Paragraphs 3-5.3 and 3-6.11. Reprinted with permission from NFPA 1002-1982, Standard for Fire Apparatus Driver/Operator Professional Qualifications, Copyright©1982, National Fire Protection Association, Quincy, Massachusetts 02269. This reprinted material is not the complete and official position of the NFPA on the referenced subject, which is represented only by the standard in its entirety.

Some basic properties and facts that apply to all relief valves are:

1. The relief valve should be set whenever more than one hose line is being supplied by the pump.
2. The relief valve should be set whenever the pump is operating in a relay operation.
3. Before the relief valve can be set, streams that are going to be used should be in operation.
4. The relief valve cannot compensate for a decrease in pressure. It only limits the amount of pressure rise that will be experienced.
5. In order for the relief valve to operate reliably, it must be exercised frequently.

Simple relief valve

In its simplest form, the automatic relief valve is a valve placed in the line to eliminate excess water (figure 1). The spring tension can be adjusted by turning the handle in or out.

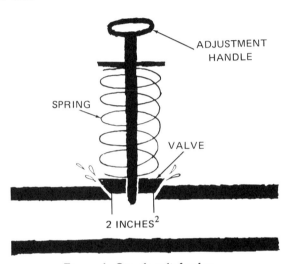

Figure 1. Simple relief valve.

As an example, the spring can be set so that 100 psi is needed to push against and open the valve. When water pressure in the pipe reaches 100 psi, the valve opens, bypassing the water and relieving the pressure. However, in order to bypass more water, it takes more pressure inside the pipe to open the valve further. The more the spring is compressed by additional pressure, the more resistance it offers.

One of the problems with this type of valve is the size required for larger pumps and the resulting spring tensions needed. For example, if the valve in figure 1 had a surface area of 2 square inches and a spring tension of 100 psi, a force of 200 pounds would be necessary to open the valve. As flows and pressure increase, bigger springs and larger valves are necessary. To overcome this deficiency, the relief valves take advantage of hydraulic action to bypass the water.

Operation through a pilot valve

By adding two hydraulic cylinders to the simple relief valve, the pressure relief system is improved (figure 2). A small pressure from hydraulic cylinder 1 will apply the same pressure to a much larger area on top of hydraulic cylinder 2. The small cylinder, known as a pilot valve, provides a big hydraulic advantage in controlling the main valve. This controlling cylinder enables the main valve to be set to bypass large volumes of water with only a slight rise in pressure, by reducing the pressure behind the piston in cylinder 2.

PUMP OPERATORS HANDBOOK

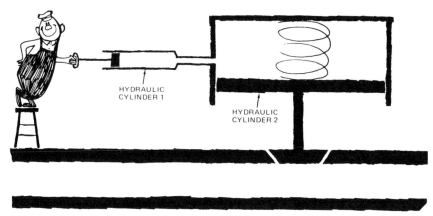

Figure 2. Pilot valve operation, step 1.

At this point control is manual. To increase pressure in cylinder 2, the piston of cylinder 1 must be pushed in. Conversely, pulling out the piston in cylinder 1 will reduce the pressure in cylinder 2. With this operation, the operator must continually monitor the control of the pilot valve.

Figure 3 illustrates the next step in the development of an automatic relief valve. A tube with a narrowing section is connected from the discharge piping to the relief valve cylinder. If the valve was left in this position, the pressure in the discharge piping would be the same as the pressure in the relief valve cylinder. However, because of the larger area of the relief valve cylinder, the relief valve should stay closed.

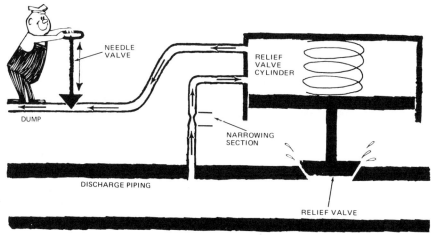

Figure 3. Pilot valve operation, step 2.

For example, if the relief valve cylinder was 4 square inches and the relief valve was 2 square inches and there was 50 psi in the discharge piping, the valve would stay closed. The force down is equal to 50×4 or 200 pounds, while the force up is equal to 50×2 or 100 pounds.

To allow the system to function, another tube is added so the water can flow out or dump. This second pipe has a needle valve on it to control the rate of dump. The needle valve can be set so that water can escape through the narrowing section.

Now, suppose the needle valve is set so that water is flowing through the narrow section to the relief valve cylinder at a rapid rate. The faster the water flows, the greater the pressure drop at the narrow section (venturi effect) and, therefore, the lower the pressure behind the relief valve cylinder.

By adjusting the needle valve setting, the hydraulic force behind the piston can be varied so the relief valve will open part way or all the way at any predetermined pressure in the pipe. This system is still manual.

PRESSURE CONTROL SYSTEMS

The spring in the relief valve cylinder only plays a small part in the operation of the relief valve. The hydraulic force behind the main valve does most of the work, and this force is controlled by the amount of water dumped.

The methods used by each of the manufacturers to automate the relief valve are covered below:

American LaFrance

The main parts of the American LaFrance relief valve are the pilot valve (13), pilot valve spring (7), adjusting screw (4), hand wheel (2), churn valve (33) and five water lines (figures 4 and 5). Note: numbers in parentheses refer to parts location in figure 4. The five water lines (figure 4) perform the following functions:

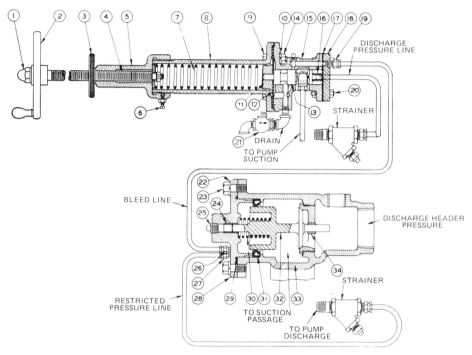

Figure 4. Cross section of an American LaFrance relief valve.

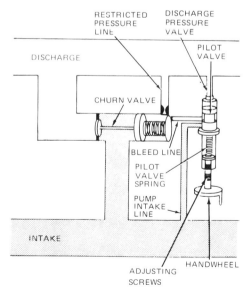

Figure 5. Simplified diagram of an American LaFrance relief valve.

1. The *discharge pressure line* runs from the pump discharge to the pilot valve housing. A strainer is included in this line to prevent foreign material from entering the pilot valve.

2. The *restricted pressure line* runs from the pump discharge to the relief valve. A strainer is included in this line to prevent foreign material from entering the relief valve or from plugging the orifice located in the tubing.

3. The *bleed line* runs from the pilot valve to the relief valve.

4. The *drain line(s)* removes water from the pilot valve and dumps it on the ground.

5. A line which connects the suction side of the pump and the pilot valve.

In normal operation, during discharge, all of the lines are filled with water and the churn valve and pilot valve are closed (figures 4 and 5). This blocks water flow from discharge to intake.

When a hose line is shut off, there is an increase in pressure in the discharge pressure line. This causes the pilot valve (13) to move, compressing the pilot valve spring (7) until the opening in the pilot valve housing is uncovered. Water is then able to flow through the bleed line and on through the pilot valve to the pump intake. This reduces the pressure on the pilot valve side of the churn valve below the discharge pressure on the other side, allowing the churn valve to move. When the churn valve opens, water is bypassed from the discharge back to the intake of the pump. The pilot valve and churn valve then equalize at the point where enough water can bypass to maintain the set pressure.

When a hose line is opened, the momentary drop in pressure causes the pilot valve to move back to the closed position, shutting the flow from discharge to intake. The pressure at the churn valve will increase from the restricted pressure line.

When the pressure in the chamber equals the discharge pressure, the churn valve will move to the closed position due to the action of the churn valve spring, thus eliminating the discharge to intake bypass. This will increase water volume and increase pressure.

The strainers in the lines between the pump discharge and the relief valve and pilot valve should be flushed periodically by opening the drain valve while the pump is operating. The valves should be closed tightly after flushing.

John Bean

The John Bean relief valve operates in a standard manner. When more than one hose line is flowing, the pilot valve spring is adjusted to maintain the desired pressure. Pressure from the discharge side of the pump keeps the churn valve seated (figure 6). Pump discharge pressure also goes to the pilot valve. This pressure is equal to the pilot valve spring pressure and the diaphragm, therefore, does not deflect. The bleed line to the churn valve remains closed.

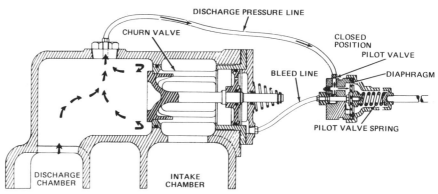

Figure 6. John Bean relief valve in closed position.

PRESSURE CONTROL SYSTEMS

Now, one of the nozzles is shut down, causing a rise in the discharge pressure. This increased pressure is transmitted to the pilot valve through the discharge pressure line, causing the diaphragm to flex as shown in figure 7. This opens the pilot valve, which allows the increased pressure to move to the churn valve via the bleed line.

Even though the pressure on both sides of the churn valve is essentially the same, the valve moves to the left because there is a larger area to the valve face on the right. When the valve moves to the left, the bypass opening between the discharge and intake is opened. Pressure is equalized so that just enough water to maintain the set pressure is bypassed.

When the nozzle is opened, the pressure causes the diaphragm to straighten, closes the bleed line, drops the pressure on the right side of the churn valve, and the combination churn valve spring and the discharge pressure close the churn valve. The bypass is thus shut off and the pressure returns to its original valve.

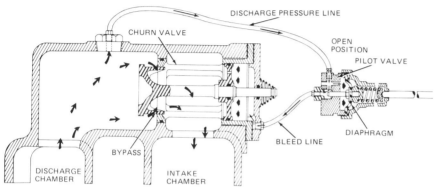

Figure 7. John Bean relief valve in open position.

Darley

The Darley relief valve is operated by turning the four-way valve handle to the ON position (figures 8, 9 and 10). The control handle can now be adjusted by turning it counterclockwise until the pressure drops about 5 psi below the set point, then slowly turning it clockwise until the gage reads the desired discharge pressure. When the four-way valve is in the OFF position, discharge

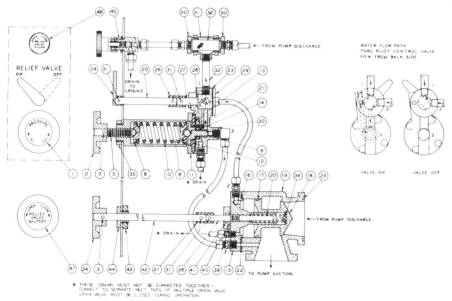

Figure 8. Schematic diagram of Darley relief valve assembly.

pressure bypasses the pilot control unit and the relief valve will not operate. Optional lights (figure 9) can indicate the status of the relief valve. (Note: The lights mentioned are not detectable in figure 9.)

This pilot control unit has a five-mesh strainer which prevents the entry of solid material. Opening the strainer flush valve periodically should remove small accumulations (figure 8 and 10). Should the water supply be so contaminated that the screen (figures 8 and 10) becomes plugged, the relief valve could stay open and prevent normal buildup of pump discharge pressure. In such an emergency, the relief valve piston can be mechanically closed by turning the relief valve shutoff control clockwise, all the way in (figure 9). Under normal operations, the relief valve shutoff must be kept fully open to allow for full travel of the relief valve.

The relief valve itself is a spring-operated valve (figures 11 and 12) which allows water to bypass from the discharge back into the intake of the pump.

Figure 9. Control panel on Darley relief valve. Optional lights indicate where it is in the open or closed position.

Figure 10. View of the Darley relief valve.

PRESSURE CONTROL SYSTEMS

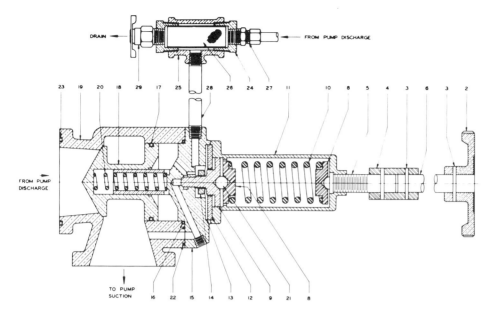

Figure 11. Schematic diagram of the Darley relief valve.

Figure 12. The Darley relief valve is spring-operated.

Hale

The Hale relief valve consists of a churn valve, a pilot valve, a pilot light and switch, and a control hand wheel (figures 13 and 14).

The churn valve is a double-ended piston with the area of the face on the pump discharge being smaller than the area on the other face.

The pilot valve contains a small rod attached to a diaphragm, a pilot valve spring, and an adjusting hand wheel. The end of the rod opposite the diaphragm is beveled and seats against an opening on the pilot valve housing. When the diaphragm is flexed by the pressure from the pump discharge, the rod moves away from the opening, allowing water, under pressure, to go the churn valve via the bleed line.

With the relief valve set and water flowing, the churn valve is closed. At this point, the pilot valve spring pressure against the diaphragm is equal to the pump discharge pressure (figure 13). When the control rod is in this position, the beveled end is seated against an opening within the pilot valve housing. The beveled end seats in such a manner that it stops the water coming through the pipe from the pump discharge and keeps it from flowing to the churn valve. The pump discharge pressure is exerted against one side of the diaphragm, while an equal pressure is exerted by the pilot valve spring on the other side of the diaphragm. Therefore, the diaphragm remains unflexed.

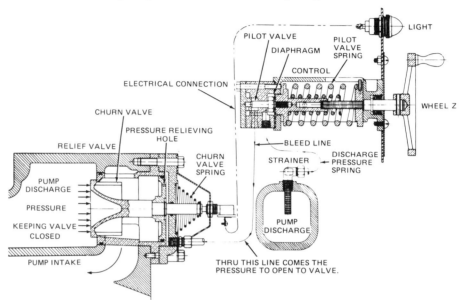

Figure 13. Schematic diagram of the Hale relief valve.

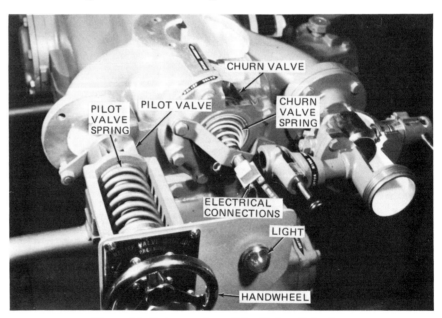

Figure 14. Cutaway view of the Hale relief valve.

As soon as the nozzle is closed, the pressure on the discharge side of the pump exceeds the pressure of the pilot valve spring and the diaphragm is flexed toward the controlling hand wheel. As the diaphragm flexes, it moves the rod away from its seat and allows water at pump discharge pressure to flow to the chamber at the large end of the churn valve.

Although water at pump discharge pressure is now acting upon both ends of the churn valve, the valve opens toward the discharge side of the pump because of the greater total force exerted on the larger face. With the churn valve open, water flows from the discharge to the intake, thereby reducing pressure.

When the discharge pressure is reduced to the point equal to the pilot spring pressure, the control rod movement stops, and the churn valve remains in a partially open position to maintain the pressure for which the relief valve is set.

When the nozzles are reopened, the pump discharge pressure acting against the diaphragm is reduced. This allows the spring pressure against the diaphragm to move the diaphragm away from the controlling hand wheel, and as the diaphragm moves, the beveled end of the rod seats against the opening from the pump discharge. This stops the flow of water to the large end of the churn valve.

The pressure retained in the water chamber at the large end of the churn valve is relieved through a small hole in the face of the piston (figure 13), allowing the pump discharge pressure and the churn valve spring to close the churn valve and stop the flow of water from the discharge to the intake. The pump pressure again equals the pilot valve spring pressure and the diaphragm is unflexed. The beveled end of the rod is then firmly seated against the opening in the line to the churn valve, thus assuming the normal operating or closed position.

The relief valve is equipped with a pilot light, activated by the pilot light switch on the churn valve (figure 13 and 14). With the churn valve in the closed position, the pilot light button is depressed by part of the churn valve assembly. When the churn valve opens, the button is released, and an electrical contact lights the lamp.

While the lamp effectively indicates that the churn valve has opened, it does not indicate the degree of the opening; hence, it cannot be used to determine if the valve is opening sufficiently to bypass the necessary amount of water. The lights remain on until the valve is completely closed.

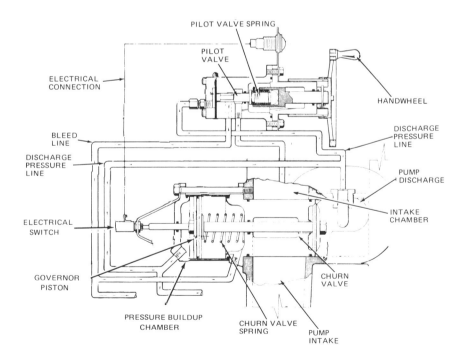

Figure 15. Schematic diagram of a Thibault relief valve.

PUMP OPERATORS HANDBOOK

Thibault

The Thibault relief valve consists of an adjusting hand wheel, a pilot valve, a churn valve, a pilot light and switch, a pilot valve spring, and a churn valve spring (figure 15).

The churn valve is a double-ended piston with the area on the left side being larger than the area on the pump intake side.

The hand wheel is adjusted to give the pilot valve spring enough tension to equal the pump pressure. Water under pressure enters the pilot valve through the discharge pressure line, and since this pressure equals the pilot valve spring pressure, the pilot valve remains closed (figure 15).

Discharge pressure is also piped to the pressure chamber of the churn valve through another discharge pressure line. From this chamber, the pressure passes through the balancing venturi and builds up discharge pressure in the piston balancing chamber (figure 15). The piston in the churn valve is now balanced and remains closed.

When a control nozzle on one of the hose lines being supplied is closed, pump discharge exceeds the pilot valve spring tension and the pilot valve opens. As the pilot valve opens, the pressure in the piston balancing chamber is reduced, because the bleed line is now connected (figure 15).

The balancing venturi is a restricted opening so the pressure goes down in the piston balancing chamber quicker than in the pressure build-up chamber. When the pressure in the pressure build-up chamber is greater than the pressure of the spring, the piston moves to the right and opens the relief valve outlet. Now, pump water can flow between the discharge and the intake and relieve the pressure. The churn valve remains in a partially open position to maintain the pressure for which the relief valve is set (figure 15).

If a nozzle is now open, the pilot valve closes, pressure builds up in both chambers, and the churn valve spring closes the churn valve, shutting off the discharge to intake opening.

The relief valve is equipped with a pilot light, activated by a pilot light switch on the churn valve (figure 15). When the churn valve opens, the switch is re-

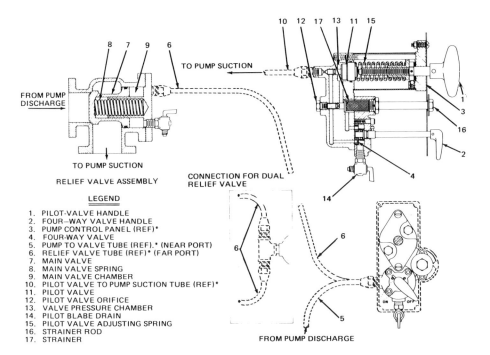

Figure 16. Schematic diagram of a Waterous relief valve.

PRESSURE CONTROL SYSTEMS

leased, turning on the pilot lamp. When the valve completely closes, the switch turns off the lamp. The lamp gives no indication of how far the churn valve has opened.

Waterous

The Waterous relief valve system consists of two units—the relief valve itself and the pilot valve that controls it (figure 16). The relief valve is spring-loaded, pressure-activated, and installed between the intake side and the discharge side of the pump (figure 17).

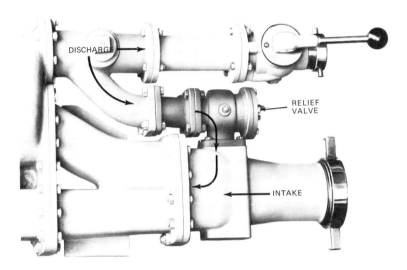

Figure 17. Waterous relief valve installed.

The relief valve opens and closes in response to directions from the pilot valve mounted on the pump control panel (figures 18 and 19).

With the pump operating, water enters the relief valve from the pump discharge manifold at full discharge pressure. Water also enters the four-way valve (4) through a tube (5) at the same discharge pressure. (Note: Numbers in parentheses refer to figure 16.) With the four-way valve on, water passes through the strainer (17) (figure 20) and fills the chamber (13) below the valve

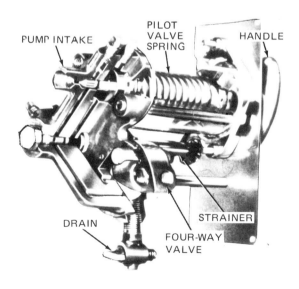

Figure 18. Rear view of the Waterous pilot valve.

Figure 19. Front view of the Waterous pilot valve.

(11). Water also passes through the orifice (12), back through the four-way valve (4), and over through the tube (6) to the relief valve chamber (9).

As long as the load applied to the valve (11) by the pump discharge pressure is less than the compression load of the spring (15), the valve remains closed, preventing discharge through the tube (10) back to the pump intake. Under

Figure 20. Waterous pilot valve strainer.

this condition, water pressure is equal on both sides of the main valve (7). Since the valve diameter is greater at the flange end of the main valve than at the seating end, the total force applied to the flange end by the water in the chamber (9) is also greater, and together with the force of the spring (8), it holds the main valve closed.

If a discharge valve is closed or if the engine is accelerated so that the pump pressure rises until the load from the pressure in the chamber (13) exceeds the

PUMP, CAB, BODY COMPONENTS

compression load from the spring (15), the valve (11) unseats. Water then escapes to the pump inlet through the tube (10). The orifice (12), through which the water must flow from the pump discharge to the relief valve, causes the pressure in the tube (6) and the chamber (9) to be lowered. The force exerted on the small end of the main valve (7) now exceeds that on the opposite end and the valve opens.

The water being pumped bypasses from discharge back to the intake side of the pump, reducing the discharge pressure. The relief valve opens just enough to reduce the discharge pressure to that set by the pilot valve.

When the discharge pressure drops below the compression settng on the spring (15), the pilot valve (11) then reseats and stops the discharge through the tube (10). Pressure then builds up in the chamber (9) behind the main valve (7) and closes it. The main valve remains closed until the discharge pressure again increases beyond the pilot valve setting, at which time the cycle is repeated.

Turning off the four-way valve (4) deactivates the pilot valve assembly. Water at discharge pressure then goes directly from the four-way valve to the relief valve, bypassing the pilot valve. The relief valve closes immediately and remains closed regardless of the discharge pressure.

Some Waterous relief valves have indicator lights at the pump panel. The lights are activated by the movement of the relief valve. A green light indicates that the relief valve is closed and an amber light shows when the valve is open.

Operation

Relief valves are placed in operation by the following procedure:

1. When two or more hand lines supplied by the pump are flowing at the volume that will be required
2. Set the discharge pressure to the amount necessary for the fireground.
3. Open the relief valve shutoff, if supplied, and set the relief valve control for the maximum possible setting. Decrease the setting until the discharge pressure gage indicates a drop or the pilot light indicates the valve has opened. Increase the setting slowly until the light goes out or the pressure gage returns to its original value. The relief valve is now set and will operate any time the pressure exceeds the set value.
4. Flush the line to the relief valve frequently if a flush control is provided. Otherwise, see that the strainer is kept clean, especially when pumping dirty water.
5. Exercise the valve frequently to ensure continuous operation.

When operating from draft or from a booster tank, it is necessary only that the relief valve bypass the same flow that is shut off, so that the torque load on the engine will be kept the same. This will keep the rpm as well as the net pump pressure constant; and since the intake pressure will not rise significantly, neither will the discharge pressure.

When operating from a hydrant connected to a high-capacity water distribution system, the intake pressure may be quite high, but it won't change significantly when the flow rate changes. Usually the net pump pressure won't be as high as when operating from draft, so less flow can be bypassed without causing friction loss through the valve system equal to the net pump pressure. This means that the system will stabilize when the water flows through the relief valve and through any open discharge lines. The system also will stabilize when the net pump pressure (equal to the friction loss through the relief valve and piping) causes a torque load equal to the torque transmitted to the pump by the engine. As indicated, there will be some increase in nozzle pressure, depending on the net pump pressure as well as the number and size of the remaining lines.

GOVERNORS

A pressure governor installed on a pumper is another way to minimize changes of pressure when volumes of water flow are changed. A governor accomplishes this by changing engine speed to compensate for changes in pressure.

The main part of the pressure governor is a cylinder (figure 21). Within the cylinder is a piston or plunger mechanically connected to the throttle linkage or carburetor of the pumper engine. The movement of the piston is under the control of the pressure being developed by the pump at its discharge and its relationship to a preset reference. This reference pressure can be established by an adjustable spring (figure 21), a pilot valve (similar to that on a relief valve), pressure in an air chamber, or hydraulic fluid.

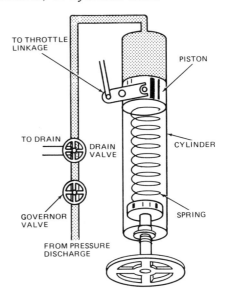

Figure 21. Typical governor.

With the governor valve opened, an increase in pump discharge pressure over reference pressure will move the throttle linkage to decrease the speed of the engine. The throttle linkage piston will keep moving in this direction until the pump rpm and the reference are equal.

If the pump pressure decreases, the piston will move in the opposite direction to increase the speed of the engine until the discharge pressure again equals the reference pressure. However, this presents an operating problem because the loss of the intake source will cause a drop in discharge pressure. The engine, in turn, will try to speed up to keep the discharge pressure constant. Since there is no intake source, the engine will speed up until it runs away. One other danger is that cavitation will take place. Cavitation, in general, results from trying to pump more water than is being supplied. (Cavitation is discussed in detail in Chapter 16.)

The drain valve is used to relieve pressure in the lines if the governor valve is closed while pump discharge pressure is still being applied. It is also necessary to drain the governor lines during cold weather to prevent water from freezing in the lines.

American LaFrance

The American LaFrance governor is composed of a balancing cylinder, a throttle linkage clutch, a reference pressure reservoir, and a control valve (figure 22). The function of each of these components is:

Balancing cylinder — transmits motion to the engine throttle linkage (figure 22).

Throttle linkage clutch — engages the balancing cylinder rod with engine throttle linkage after reference pressure has been set. If pump discharge pressure falls below 50 psi, the clutch will disengage to prevent running away and cavitation. The clutch is spring-loaded to disengage completely when not pressurized to allow free movement of the throttle arm during road operation (figure 23).

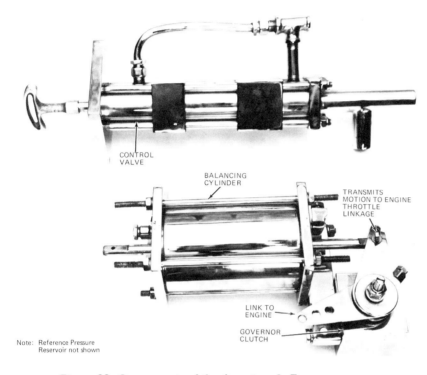

Figure 22. Components of the American LaFrance governor.

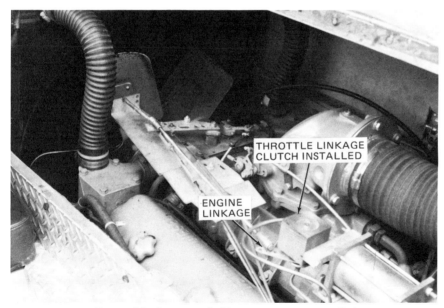

Figure 23. American LaFrance governor installed.

Reference pressure reservoir — accumulates the pressure desired from the fire pump, which is then trapped between the air in the top of the cylinder and the reference side of the piston in the balancing cylinder (figure 25). The air compressed in the top of the cylinder acts as a spring to impart movement to

PUMP OPERATORS HANDBOOK

the piston and to maintain a balance of reference and pump pressures. An increase in the engine speed occurs when the volume of air is increased. This volume of air also allows a decrease to the amount of pump pressure set initially, with a decrease in discharge volume.

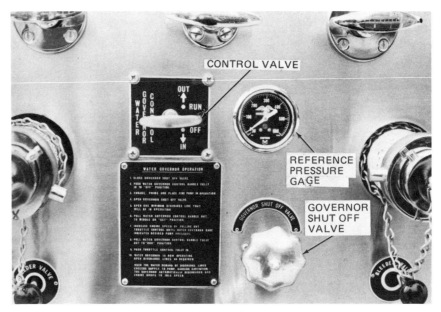

Figure 24. American LaFrance governor operating panel.

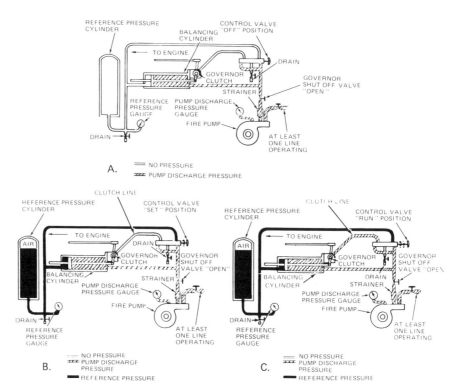

Figure 25. Schematic diagram of the American LaFrance governor.

Control valve — controls the operation of the governor system at the pump panel (figure 22).

The governor systems operates as follows:
1. The pump is placed in operation in the normal manner.

PRESSURE CONTROL SYSTEMS

2. Open the smallest discharge line that will be used and advance the hand throttle to obtain a slight pressure. Now, when the clutch is engaged, the balancing piston is at the end of its travel, at the slowest possible setting.

3. Open the governor shutoff valve (figures 24 and 25A). Pump discharge pressure is now fed to one side of the balancing cylinder.

4. Move the control valve to the set position (figures 24 and 25B). This pressurizes the line through the set circuit (solid area in the illustration) to the reference pressure reservoir and to the reference pressure side of the piston in the balancing cylinder.

5. Advance the hand throttle until the desired pressure is obtained on the reference pressure gage (figures 24 and 25B). This raises the pressure in the entire governor system, with the exception of the clutch line.

6. Move the control valve to run and push the hand throttle closed (figures 24 and 25C). Pump discharge pressure is now connected to the clutch through the clutch line. This engages the throttle linkage with the balancing cylinder rod. The hand throttle is closed so that if pump pressure drops below 50 psi, the engine will drop to idle speed when the clutch disengages.

7. When additional lines are placed in service, pressure will drop in the pump discharge side of the balancing cylinder. The higher reference pressure will move the cylinder to the right (figure 25C), increasing engine speed.

8. To shut down the operation, return the control valve to the off position, open the reference pressure cylinder drain, close the governor shutoff valve, and open the control valve drain (figure 25B).

9. There is a strainer in the line between the pump and the control valve. This strainer must be checked periodically to ensure that it is not clogged.

American

The American governor consists of an assembly that fits on the carburetor, a control rod connected to a piston in a water chamber, and a control valve (figure 26).

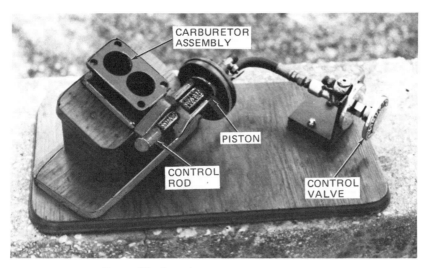

Figure 26. American governor components.

The governor operates as follows:

1. Open the control valve all the way. This will allow water to be discharged faster than it can flow into the water chamber through the restrictor in the line and keep the governor from operating (figure 27).

2. Hose lines should be established and water should be flowing at the desired pressure and quantity.

3. Check the position of the governor return valve (figure 27). This valve

allows the water to either return to the pump intake or dump on the ground through the overflow pipe. When operating from a pressure source, the valve should be opened, but it must be closed when operating from draft. If the valve is left open when drafting, it will be impossible to establish a prime since there would be air entering the pump through the overflow pipe. On the other hand, if the valve is left closed when the pump is fed from a pressure source, the governor will not operate consistently since the change in pressure in the intake will change the flow resistance from the return line. This, in turn, will change the rate of escape for the water and thus the operating value of the governor.

4. The control valve now is closed slowly until the pressure on the discharge gage begins to drop. The valve now should be opened until the pressure readings return to the desired value. Now, if a line is shut down, the increased pressure on the diaphragm of the piston assembly exceeds the force of the spring, and the control slows down the engine speed (figure 27).

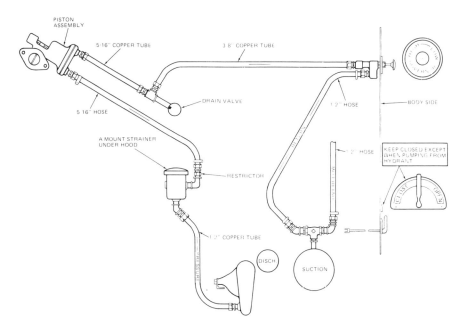

Figure 27. Schematic diagram of the American governor.

5. If the line is now opened, the intake pressure drops and the spring moves the control rod to increase engine speed.

6. If the control is kept completely closed, then it will not be possible to develop more than about 110 psi discharge pressure.

7. There is a strainer in the line between the pump discharge and the piston assembly (figure 27). This strainer should be cleaned periodically, especially after pumping dirty water.

8. To prevent freezing, the drain valve should be opened after use of the governor during cold weather.

Hale

The Hale governor consists of an actuator, throttle, dampener needle, the air tank, and the carburetor linkage (figure 28). The governor operates as follows:

1. Hose lines should be open and the water flowing at the desired pressure and quantity. Pressure is increased by backing out the throttle knob to obtain about 10 psi above the desired operating pressure.

2. Wait approximately three seconds to allow the air tank to fill to operating pressure. Then, pull out the actuator all the way to engage the O-ring seal

PRESSURE CONTROL SYSTEMS

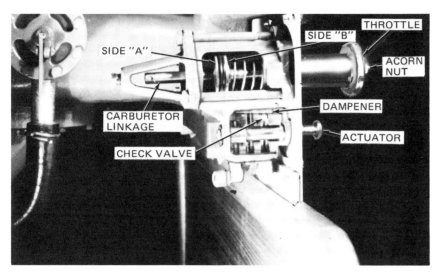

Figure 28. Components of the Hale governor.

(figure 29). This now seals off the pump discharge pressure so that it only pushes against side "B" of the piston (figure 29). At the same time, the air pressure is trapped in the tank, resulting in a constant reference pressure pushing on side "A" of the piston (figures 28 and 29).

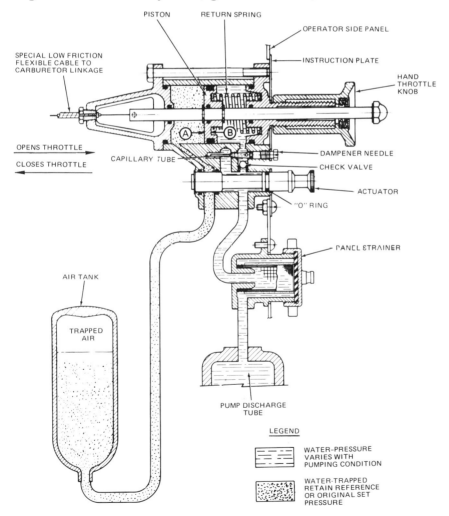

Figure 29. Schematic diagram of the Hale governor.

3. Turn the throttle knob all the way in. The acorn nut throttle linkage will remain out (figure 29).

4. If a line is shut down, an increase in pressure will raise the pressure on face "B" of the piston (figures 28 and 29). This will close the throttle.

5. The check valve and adjustable dampener needle valve (figures 28 and 29) minimize the tendency of the piston to override or go into a surging condition. The check valve opens on rising pump pressure to permit a fast response of the piston in closing the throttle. When the pump pressure drops (from the opening of a line for example), the check valve closes and the water leaving the cylinder is diverted through and around the needle valve. The needle valve contains a short length of capillary tubing which slows down the rate of throttle opening and prevents surging.

6. If the throttle knob is difficult to turn, the dampener needle can be removed and the capillary tube checked for dirt. Be sure to relieve pressure completely before removing the dampener needle, or the check valve ball may be blown out the dampener needle hole.

7. If more than three seconds are required for the air tank to fill to operating pressure, the strainer may be clogged. Remove the strainer cap and pull out the strainer screen (figure 29). Flush the metal screen and polyethylene tubing filter until clear.

8. During freezing weather, the governor and piping require draining. Open the drain cock on the bottom of the air tank to drain the tank. With the main pump drains open, move the throttle to its extreme position to help drain the governor body. Close all drains when finished.

Seagrave

The Seagrave governor is composed of two major assemblies, the hydraulic remote control and the piston assembly (figures 30 and 31). The governor operates as follows:

1. Hose lines should be opened and the water flowing at the desired pressure and quantity. Pressure is increased by opening the throttle at least 10 psi above the desired pressure. The governor control handle should be set at its highest value (figures 30 and 32).

2. Open the governor valve (figures 30 and 32). This permits water under discharge pressure to enter the piston assembly and exert a pressure on one side of the diaphragm.

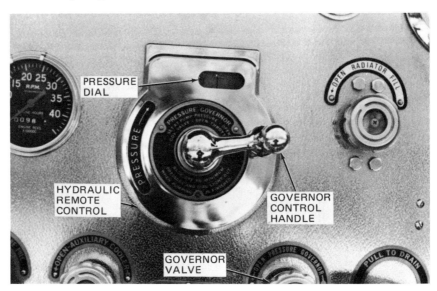

Figure 30. Seagrave governor hydraulic remote control.

PRESSURE CONTROL SYSTEMS

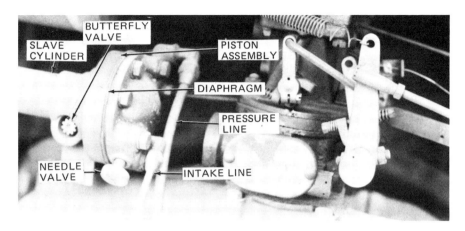

Figure 31. Seagrave governor piston assembly.

3. Lower the governor setting by slowly turning the dial. Watch the discharge gage and continue to turn the control handle until the gage reads the desired discharge pressure. Now, open the throttle a few turns above the original setting (figures 30 and 32). Since the hydraulic remote control consists principally of a small automobile brake master cylinder, moving the handle in forces the piston to compress the spring. This forces the hydraulic fluid out through the tubing line and into the hydraulic cylinder of the piston assembly (figure 32). This forces the piston in the slave cylinder forward, compressing the spring and exerting a pressure on the diaphragm.

4. When a hose line is shut down, the intake pressure increases. This exerts a pressure on the diaphragm that is greater than the pressure from the slave cylinder. The diaphragm then moves, closing the butterfly valve which slows the engine down (figures 31 and 32).

5. The needle valve (figures 31 and 32) has a flattened end to permit a slight, continuous flow of water back to the intake. This permits the governor to stabilize and reduce the surging action.

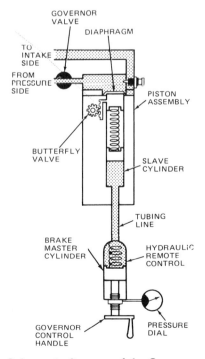

Figure 32. Schematic diagram of the Seagrave governor.

6. If the flow through the needle valve should stop due to clogging, the obstruction can be flushed out by opening the needle valve a few turns for a few seconds and reclosing.

7. The water-operated parts of the governor are drained when the pump is drained. The connection from the remote control to the piston assembly is accomplished with hydraulic fluid, so there is no freezing problem.

Waterous

A mechanical-hydraulic engine governor maintains constant pump pressure at a desired setting by adjusting the engine throttle (figures 33 and 34). The system consists of a panel-mounted directional flow valve and an on-off valve which, together with a bladder-type accumulator, control a hydraulic cylinder connected to the engine throttle linkage (controls to 500 psig).

Three major units comprise the pressure controller system: the directional flow valve assembly, the actuating cylinder assembly, and an accumulator. The panel control (figure 33) is used to adjust the system pressure and to place the controller into the out-of-service mode.

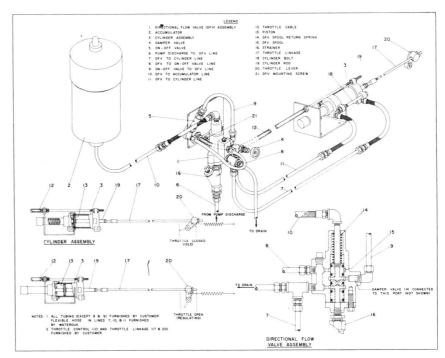

Figure 33. Schematic diagram of the Waterous governor.

To activate the pressure controller system:

1. Accelerate the engine with the hand throttle until the desired discharge pressure is obtained.
2. Turn the pressure controller on.
3. Pull the hand throttle all the way out.
4. If the pump loses prime or the intake pressure drops below 30 psi, the controller cuts the engine speed to an idle.

The pressure controller system operates as follows:

1. The actuator cylinder assembly, connected to the engine throttle, controls engine speed in response to the reference pressure.
2. The large, precharged, bladder-type accumulator establishes the reference pressure for the pressure controller system. When the system is adjusted, the reference pressure is stored in the accumulator.
3. As pump discharge pressure fluctuates above or below the set pressure,

PRESSURE CONTROL SYSTEMS

the reference pressure in the accumulator works through the cylinder to increase or decrease engine speed to compensate for the pressure variation.

4. When the accumulator is installed, it is precharged to 75 psi with air or nitrogen to provide closer pressure control when the pump pressures change. A bladder separates the gas from the water in the accumulator to prevent the system from becoming waterlogged.

Figure 34. Directional control valve assembly mounted on pump control panel with hand throttle and accumulator.

AUTOMATIC GOVERNOR

A new device for automatic governor operation has been recently developed. This automatic governor consists of a pump panel mounted control, a fuel feed control, and a remote control (figures 35 and 36). The governor operates as follows:

1. Hose lines should be open and the water flowing at the desired pressure and quantity. The governor control switch is set off (figure 35).

2. The governor control switch can now be set to manual and the discharge pressure can be regulated by pressing the increase or decrease push switch un-

Figure 35. Automatic governor pump panel control.

til the new discharge pressure is obtained. The unit will then maintain this new pressure (figure 35).

3. If the governor control switch is set to auto, the right-hand pressure switch can be set to the desired discharge pressure and the governor will automatically maintain the set pressure (figure 35).

4. When the governor control switch is set to remote, the governor can be operated from a point near the fire rather than from the truck (figure 35).

5. The controls are all solid state devices (no tubes) so that freezing and heating action due to hydraulic changes are avoided (figure 36).

Figure 36. Automatic governor fuel feed control.

Chapter 14

Priming Devices

A pump operator often has to use a static water source, such as a pond or lake, to supply water to the fireground. However, since water cannot be pulled or lifted, it is necessary to push water into the pump with some kind of pressure.

Under static conditions, the atmospheric pressure being exerted on the surface of the open body of water is used to push the water into the pump. This is done by creating an airtight waterway from the surface of the water into the pump, and then lowering the pressure within the pump and waterway. In this way, the atmospheric pressure on the outside of the pump is greater than the pressure on the inside and the water is pushed into the pump.

When the water has risen to a height within the passageway so that the pressures are balanced, the water will stop flowing. When the amount of vacuum created is sufficient to overcome the pounds per inch back pressure created by the elevation of the water, a flow can be established. The establishment of this vacuum, then, is the object of the priming system.

"The fire apparatus driver/operator shall identify the theory and principles of pumper priming systems."*

It is important to remember that only centrifugal pumps must be primed in order to pump water from a static source. Priming is necessary because a centrifugal pump has a continuous waterway from the intake to discharge and, therefore, cannot pump air. However, once primed, the movement of water through the pump will exclude air and maintain the vacuum.

On the other hand, a positive displacement pump needs no external priming device. The movement of this type of pump is such that it will pump air as well as water. If an airtight waterway can be established from the intake to the water surface, air will be discharged. Since no air can enter, a vacuum will be established and water will enter to replace the discharged air.

The three general methods for priming a centrifugal pump are by use of:
1. Positive displacement pump — rotary gear or rotary vane type
2. Exhaust pump — venturi from the exhaust manifold
3. Vacuum pump — suction from the intake manifold.

ROTARY PRIMING PUMPS

Since the positive displacement pump is self-priming, it can purge itself of air. It can be connected to a centrifugal pump to provide the vacuum that cen-

---

*Paragraph 3-5.2. Reprinted with permission from NFPA 1002-1982, Standard for Fire Apparatus Driver/Operator Professional Qualifications, Copyright© 1982, National Fire Protection Association, Quincy, Massachusetts 02269. This reprinted material is not the complete and official position of the NFPA on the referenced subject, which is represented only by the standard in its entirety.

trifugal pumps are unable to accomplish alone. In this case, the intake side of the priming pump is connected through a valve to an opening in the pump casing and the discharge of the priming pump is usually piped to a drain, discharging on the ground. In most applications, either a rotary gear or rotary vane type of pump is used for priming (figure 1).

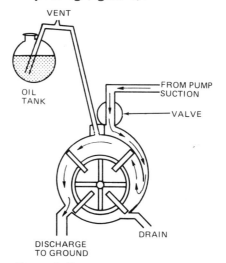

Figure 1. Rotary vane priming pump.

These primers use an oil supply to seal and lubricate the gears more efficiently and to improve the priming capabilities. The vent in the oil line breaks the vacuum when priming is complete, so that the oil will not be siphoned out of the tank.

The primers are usually driven either mechanically from the transfer case or by an electric motor. If an electric motor is provided, it is common practice to provide a mechanical arrangement for emergency usage.

Operation of the rotary type primer varies from pump to pump, but the mechanical type generally operates from an engine speed of 700 to 1500 rpm. The electric type runs independently of engine speed. With either type, when the prime is accomplished, water will be discharged from the drain piping.

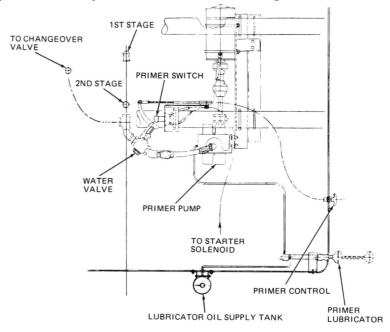

Figure 2. Schematic diagram of the American LaFrance electric priming system.

PRIMING DEVICES

Although this type of primer does not require much maintenance, it is absolutely essential to keep the oil reservoir filled at all times.

Some basic principles that apply to all rotary type primers are:

1. If the connection from the primer is made to the intake side of the pump near the impeller, the pump should be turning while it is being primed.

2. If the connection from the primer is made on top of the pump, the impellers should not be turning when it is being primed.

3. The primer should be capable of providing 22 inches of mercury reading on the compound gage within 30 seconds. Mercury is heavier than water and a 1-inch column of mercury (Hg) is the equivalent of 1.31 feet of water column. A reading of 22 inches Hg would therefore indicate a negative pressure in the pump capable of a static lift of 24.86 feet of water ($22 \times 1.13 = 24.86$ feet).

American LaFrance

The major components of the American LaFrance primer are the lubricator oil supply tank, the primer control, the primer pump, the water valve, and the primer switch (figure 2).

The primer operates as follows:

1. The primer control handle is pulled out from the pump panel. The cable control opens the water valve which opens the passageway between the main

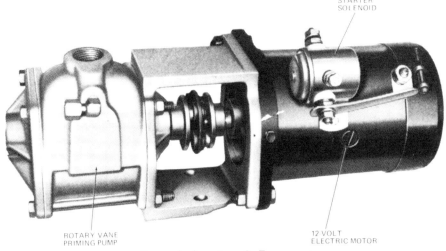

Figure 3. American LaFrance primer.

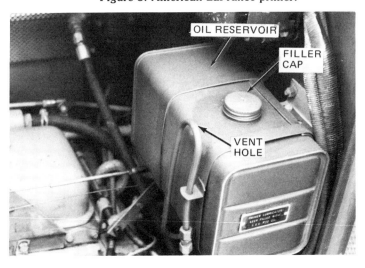

Figure 4. American LaFrance oil reservoir.

pump and the rotary vane priming pump (figure 2). Pulling the control handle also activates the priming switch.

2. With the priming switch on, the 12-volt motor turns the rotary vanes of the primer pump (figure 3).

3. The rotary vane pump creates a vacuum and draws air out through the water valve from the main pump.

4. When all the air has been removed from the pump, water will flow from the primer out to the ground.

5. Oil from the reservoir is siphoned to the rotary vane pump (figure 4.) When priming is stopped, the vacuum in the oil line is broken by the vent hole. In some models, oil from the reservoir is supplied by manually operating the primer lubricator control (figure 2).

Darley

The Darley electric priming pump uses a rotary vane primer (figure 5). This pump's main components are the lubricator oil supply tank, the primer control, the primer pump and the water control valve (figure 5).

The primer operates as follows:

1. The primer control handle is pulled out. This activates the switch that turns on the power to the motor and opens the passageway between the main pump and the rotary vane pump.

2. The motor turns the rotary vane pump which draws air out through the water valve from the main pump.

3. When all the air has been removed from the pump, water will flow out onto the ground through the primer exhaust pipe.

4. Oil from the reservoir (figure 5) is siphoned to the rotary vane pump. When pumping is stopped, the vacuum in the oil line is broken by the vent hole (figure 5). The oil reservoir uses SAE 30 motor oil and has a capacity of 6 quarts.

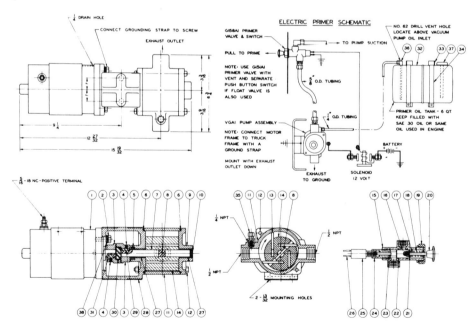

Figure 5. Schematic diagram of the Darley electric priming pump.

Hale

Hale pumps are equipped with either a vacuum/air system or an electric priming system. The major components of these systems are the priming

PRIMING DEVICES

pump, the priming valve, the priming clutch (vacuum/air operation only), and the control source (figures 6 and 7).

The vacuum/air priming system operates as follows:

1. When the priming button (figure 8) is pushed in, the area on the inside of the priming valve diaphragm and the outside of the priming clutch diaphragm are subjected to vacuum from the engine intake manifold (figure 6).

2. In the priming valve, atmospheric pressure on the outside of the diaphragm pushes the valve open, connecting the priming pump suction to the main pump intake (figures 6 and 7).

3. The priming clutch is engaged by atmospheric pressure pushing on the inside of the priming clutch diaphragm, thus driving the gears of the rotary priming pump (figure 6). Levers are provided on the priming valve and priming pump clutch for emergency manual operation (figures 6, 8 and 9).

4. When the priming clutch is engaged, the priming pump (figure 10) is driven by a gear in the pump gear case. The speed of the priming pump will be dependent on the speed of the pump motor. Most pumps are designed for the

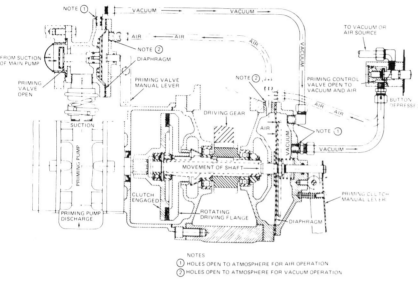

Figure 6. Schematic diagram of the Hale vacuum/air primer.

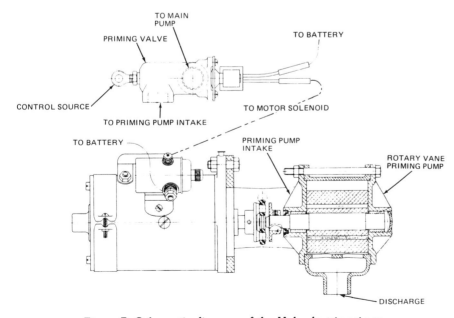

Figure 7. Schematic diagram of the Hale electric primer.

motor to be turning approximately 120 rpm during the priming operation. The priming pump on the Hale pump will not turn unless the main pump is engaged and the road gear is disengaged.

5. The mechanical operation of the priming valve and the priming pump will be the same for air pressure operation, except that the air will be connected to the opposite side of the diaphragm (figure 6).

6. The priming pump uses a supply of motor oil to seal the gears and provide lubrication for the pump.

The electric priming system operates as follows:

1. Pulling the control handle opens the priming valve, which connects the main pump intake to the rotary gear primer (figure 7) and activates the electric motor.

2. As the rotor turns, it creates a vacuum on the intake side of the main pump. At the same time, oil is pulled in from the oil tank for lubricating and sealing the priming pump.

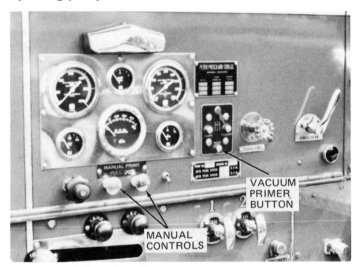

Figure 8. Hale vacuum primer controls.

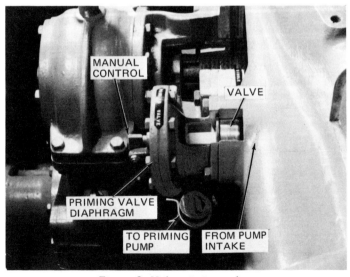

Figure 9. Hale priming valve.

Seagrave

The Seagrave rotary vane priming pump is driven by a clutch arrangement when the control rod is pulled (figure 11).

PRIMING DEVICES

The operation sequence for the pump is:
1. When the priming pump is disengaged, the cone clutch is not connected (figure 12). Now, when the control rod is pulled, the clutch is engaged and the pump gears drive the rotary vane priming pump (figures 11 and 13).
2. With the rotary vane pump turning, air is pumped out of the main pump.
3. An oil reservoir keeps the vanes lubricated and helps to seal the pump.

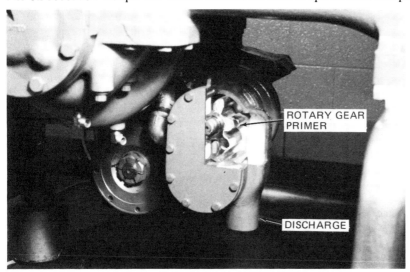

Figure 10. Hale rotary gear primer.

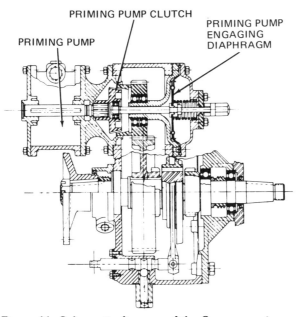

Figure 11. Schematic diagram of the Seagrave primer.

Waterous

The major components of the Waterous primer are the priming valve, rotary gear priming pump, and control rod (figure 14).

The primer operates as follows:
1. Pulling (or pushing, depending on the particular installation) the control rod opens the priming valve, connecting the rotary gear primer to the first-stage intake chamber of the main pump (figure 15). At the same time, pulling the rod activates the electric switch which turns on the priming motor (figure 16).

PUMP OPERATORS HANDBOOK

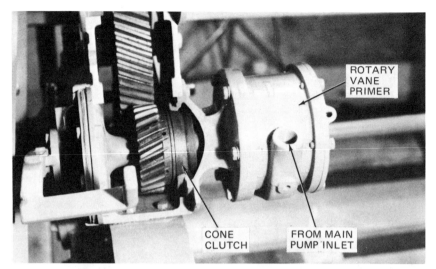

Figure 12. Seagrave primer with clutch disengaged.

Figure 13. Seagrave primer with clutch engaged.

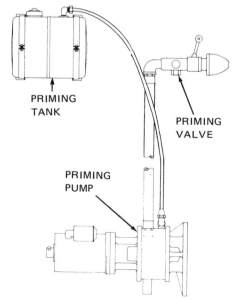

Figure 14. Schematic diagram of the Waterous primer.

PRIMING DEVICES

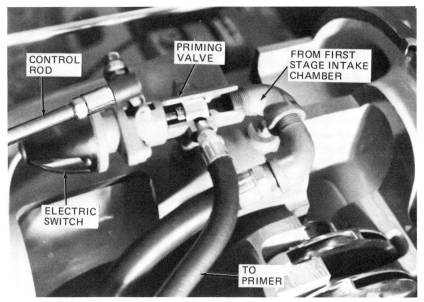

Figure 15. Waterous priming valve.

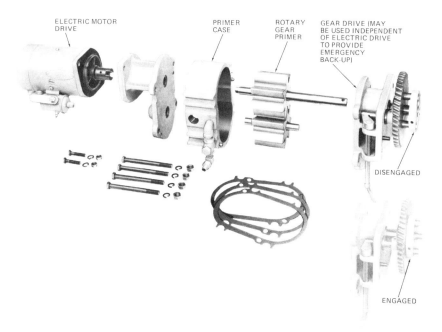

Figure 16. Waterous primer and drive assembly.

2. The priming motor turns the rotary gears, removing air from the intake side of the main pump and discharging it from the rotary gear pump.

3. Oil from the auxiliary priming tank (figure 14) is siphoned into the rotary gear pump to provide an airtight seal and to lubricate the rotors.

4. If the primer is equipped with an electric drive, a manual shift for engaging the priming motor with the pump drive can be provided (figures 16 and 17). Then, if the electric motor should fail, the pump primer can be engaged manually.

VACUUM PRIMER

One of the simplest types of primers, the vacuum primer, makes use of the fact that any gasoline-powered engine creates a vacuum at its intake manifold

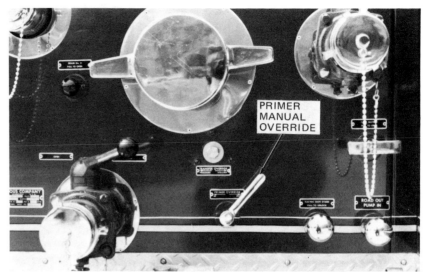

Figure 17. Waterous manual primer override.

at all times when it is running. In order to take advantage of the vacuum, a device can be connected between the pump and the intake manifold so that a connection can be made when necessary. To do this, the priming device must include safeguards to prevent two inherent dangers:

1. Water must be prevented from being drawn into the engine.
2. Gases from a backfire must be prevented from being forced into the pump.

The American and Darley pumps use a vacuum primer which operates as follows:

1. Pulling the priming handle connects both the intake manifold and the pump intake to the priming assembly (figures 18 and 19).
2. The vacuum from the engine draws air from the pump intake, through the main float area, past the lower ball valve, past the safety float and valve, out through the pressure valve to the engine manifold (figures 18, 19, 20 and 21).
3. When a sufficient vacuum has been created, water enters the main float area and raises the float. This, in turn, raises the lower ball valve to shut off the vacuum to the pump intake. The upper ball valve is also raised so that outside air goes to the engine manifold (figures 11, 17, 20 and 21).
4. If the main float should fail to rise, the lower ball valve will remain open. Water then will rise into the upper chamber, raising the safety float and closing the safety valve. This seals off the engine manifold and prevents the water from being drawn into the engine (figures 18, 19, 21 and 22).
5. If the engine should backfire while the primer is operating, the increased pressure will cause the pressure valve to close, thus shutting off the primer from the engine (figure 18).
6. When the priming handle is released, the engine manifold is disconnected from the upper chamber and the lower chamber is connected to the drain. Now, water in the lower chamber can be drained to prevent freezing and to lower the main float (figure 18).
7. The maximum vacuum at the intake manifold is created when the engine is operating at an idle, approximately 800 rpm, with as little load on the engine as possible.
8. These vacuum primers should be checked frequently. The main float, linkage, and ball valves should be checked for wear or damage. The cork safety float should be checked for deterioration. This is especially important because a failure will result in water getting into the engine.

PRIMING DEVICES

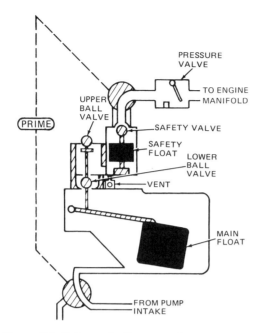

Figure 18. Schematic diagram of a vacuum primer.

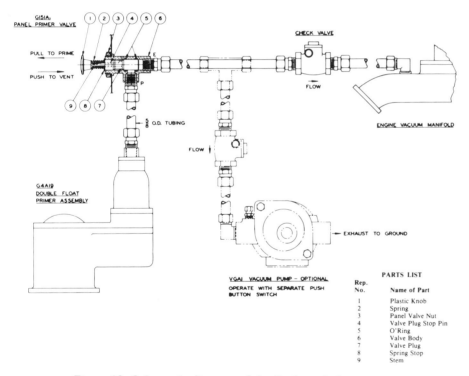

Figure 19. Schematic diagram of the Darley priming pump.

EXHAUST PRIMER

The exhaust primer makes use of the venturi principle. The rapidly moving exhaust gases from the apparatus engine are diverted through a chamber which connects to an opening in the pump housing. The rapidly moving gases tend to draw the air out of the pump, creating a vacuum.

This particular type primer requires a very high engine rpm because the faster the gases are moving, the more pronounced will be the action of the primer in exhausting the air.

179

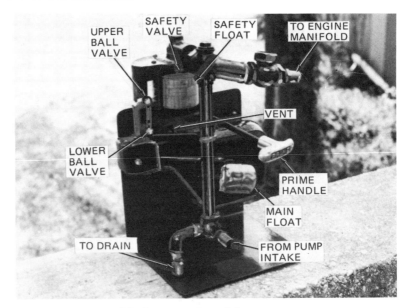

Figure 20. Cutaway view of the vacuum primer.

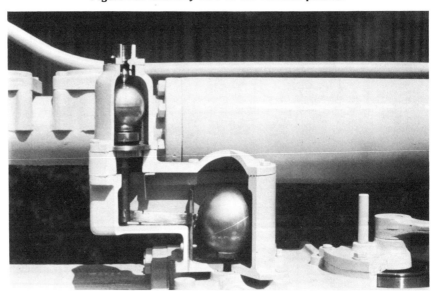

Figure 21. Darley double float primer.

The exhaust primer operates as follows:

1. Pulling the primer handle closes the butterfly valve, which deflects the exhaust gases through the exhaust venturi, and opens the priming valve which connects the pump intake to the primer (figures 22 and 24).

2. The exhaust gas is diverted through the venturi, reducing the pressure in the line from the pump intake. The reduced pressure opens check valve "E" and air from the pump intake flows to the ground (figure 23).

3. The vacuum created in the line keeps check valve "W" closed.

4. If the priming valve should be accidentally opened while connected to a pressure source, water would enter the exhaust venturi. To prevent this, the water venturi is installed. Water under pressure, passing through the venturi, will create a vacuum, thus closing check valve "E." Check valve "W" would be forced open, thus dumping water on the ground (figure 23).

5. Since this type of primer uses exhaust gases with their carbon deposits, water, and other waste products, it is more in need of maintenance than some of the other primers. The condition of the truck's exhaust system is critical to

PRIMING DEVICES

the operation of this system. The primer's chamber also requires cleaning at intervals. Operation of some of the valves that are used in diverting the exhaust gases depends on frequent lubrication and maintenance.

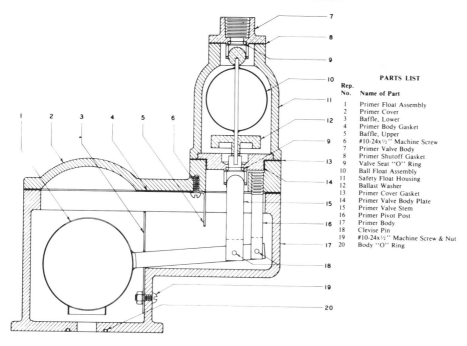

Figure 22. Schematic diagram of the Darley double float primer.

PARTS LIST

Rep. No.	Name of Part
1	Primer Float Assembly
2	Primer Cover
3	Baffle, Lower
4	Primer Body Gasket
5	Baffle, Upper
6	#10-24x½" Machine Screw
7	Primer Valve Body
8	Primer Shutoff Gasket
9	Valve Seat "O" Ring
10	Ball Float Assembly
11	Safety Float Housing
12	Ballast Washer
13	Primer Cover Gasket
14	Primer Valve Body Plate
15	Primer Valve Stem
16	Primer Pivot Post
17	Primer Body
18	Clevise Pin
19	#10-24x½" Machine Screw & Nut
20	Body "O" Ring

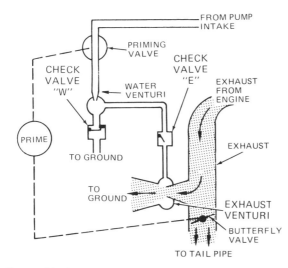

Figure 23. Schematic diagram of an exhaust primer.

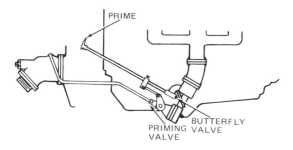

Figure 24. Thibault exhaust primer.

181

Chapter 15

Drafting Operations

Drafting water is the process of using water from a static source, such as a pond, lake or basin. Since the water source is static, or at rest, the pump operator's job is to have the water forced into the pump so that it can be delivered to the fireground.

"The fire apparatus driver/operator, given a fire department pumper, shall identify the principle of drafting water, and demonstrate a systems check when the pumper will not draft."*

As explained in Chapter 14, each centrifugal pump is a priming device to eliminate air from the pump and intake lines. Once the air has been eliminated and a partial vacuum created, atmospheric pressure pushes water into the pump (figure 1) through the noncollapsible (hard suction) hose.

Lifting water

"The fire apparatus driver/operator, given the necessary information, shall compute the maximum lift of a fire department pumper."**

For each 1 inch of mercury vacuum created, water will be pushed into the noncollapsible hose (or hard sleeve) a distance of 1.13 feet. Since lift is measured from the surface of the static source to the center line of the pump, a perfect vacuum in the pump, at sea level, will allow water to be pushed to a height of

$$14.7 \text{ psi} \times 2.304 \text{ ft/lb} = 33.86 \text{ ft, or}$$
$$29.92 \text{ in of mercury} \times 1.13 \text{ ft/in} = 33.81 \text{ ft}$$

The perfect vacuum necessary for the theoretical lift of 33.9 feet is almost impossible to achieve even in a laboratory. Additional loss in optimum lift is accounted for due to friction in the suction hoses (a pump may lift 500 gpm through a 5-inch suction; however, the same pump may lift 500 gpm only 12½ feet when a 3½-inch suction is used). Head loss, water temperature, atmospheric pressure at the site location and condition of the pump all contribute to lessening the theoretical height that a pumper may lift water. This limits practical lifts to 28 feet for an excellent rating; 25 feet for a good rating. Most pumpers in service lift somewhat less than these figures.

The height of the lift is independent of the angle of the hard sleeve. The lift in both situations of figure 2 is the same. The only distance of importance, as far

---

*Paragraph 3-6.3. Reprinted with permission from NFPA 1002-1982, Standard for Fire Apparatus Driver/Operator Professional Qualifications, Copyright©1982, National Fire Protection Association, Quincy, Massachusetts 02269. This reprinted material is not the complete and official position of the NFPA on the referenced subject, which is represented only by the standard in its entirety.
**Paragraph 3-4.11. Ibid.

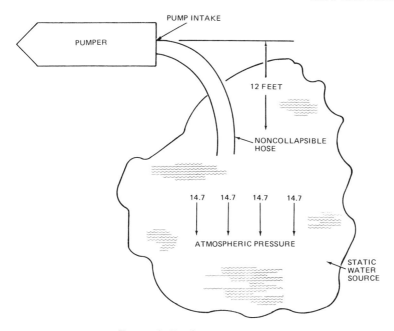

Figure 1. Drafting operations.

as lift is concerned, is the vertical distance. However, longer hard sleeves require additional work by the pump to overcome the increased friction loss.

Since the vacuum gage is calibrated in inches of mercury, the atmospheric pressure is measured in psi, and the lift is calculated in feet (table 1 provides the conversions between units). Using the table, the 12-foot lift shown in figure 1 requires a vacuum of 10.60 inches of mercury or 5.20 psi. This means that the

TABLE 1. Units of Pressure Conversion Table

Feet of Water (Lift)	Inches of Mercury	Psi	Feet of Water (Lift)	Inches of Mercury	Psi
7.00	6.18	3.03	20.79	18.36	9.00
7.93	7.00	3.43	21.00	18.54	9.09
8.00	7.06	3.46	21.53	19.00	9.32
9.00	7.95	3.90	22.00	19.43	9.53
9.06	8.00	3.92	22.66	20.00	9.81
9.24	8.16	4.00	23.00	20.31	9.96
10.00	8.83	4.33	23.10	20.40	10.00
10.20	9.00	4.42	23.79	21.00	10.30
11.00	9.71	4.76	24.00	21.19	10.39
11.33	10.00	4.91	24.93	22.002	10.79
11.55	10.20	5.00	25.00	22.08	10.83
12.00	10.60	5.20	25.41	22.44	11.00
12.46	11.00	5.40	26.00	22.96	11.26
13.00	11.48	5.63	26.06	23.00	11.28
13.60	12.00	5.89	27.00	23.84	11.69
13.86	12.24	6.00	27.19	24.00	11.77
14.00	12.36	6.06	27.72	24.48	12.00
14.73	13.00	6.38	28.00	24.72	12.12
15.00	13.25	6.50	28.33	25.00	12.27
15.86	14.00	6.87	29.00	25.61	12.56
16.00	14.13	6.93	29.46	26.00	12.76
16.17	14.28	7.00	30.00	26.49	12.99
17.00	15.00	7.36	30.03	26.52	13.00
18.00	15.89	7.79	30.59	27.00	13.25
18.13	16.00	7.85	31.00	27.37	13.42
18.48	16.32	8.00	31.72	28.00	13.74
19.00	16.78	8.23	32.00	28.76	13.86
19.26	17.00	8.34	32.34	38.56	14.00
20.00	17.66	8.66	32.86	29.00	14.23
20.39	18.00	8.83	33.00	29.14	14.29
			33.90	29.92	14.70

partial pressure in the pump can be

$$14.7 - 5.2 = 9.5 \text{ psi}$$

and the pump will still lift the water.

The partial vacuum required under static conditions, with no water being discharged, is 5.2 psi. It is the minimum vacuum required to do the job of letting water be pushed into the pump. Once the water begins moving, a higher vacuum will be necessary to overcome the friction loss in the intake hose.

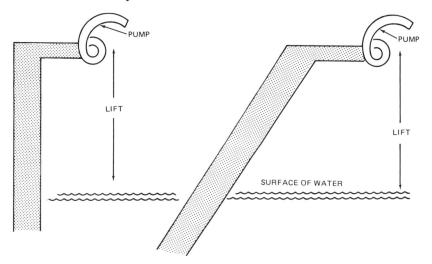

Figure 2. Measuring lift.

Climatic conditions

Performance of a pump also depends on the climate for drafting operations. Since the ability to lift water depends on atmospheric pressure, the specific pressure will change with the weather. On clear, fair days atmospheric pressure is higher than on cloudy or stormy days. The pressure change will therefore influence the maximum lift.

Even more important than the small weather variation of atmospheric pressure is the variation due to a change in altitude. The pressure drops approximately 1 inch of mercury (½ psi) for every 1000 feet of altitude above sea level. Table 2 provides atmospheric pressure for various altitudes.

Normally, the temperature of the water available for drafting is not warm enough to cause concern. However, if water is being used from a booster tank or a test pit, the temperature can have a very noticeable effect. Every body of

TABLE 2. Atmospheric Pressure Versus Altitude

Altitude (feet)	Pressure (psi)	Loss of Lift Feet of Water	Altitude (feet)	Pressure (psi)	Loss of Lift Feet of Water
−1000	15.2	1.16	4500	12.4	5.3
−500	15.0		5000	12.2	5.8
0	14.7		5500	12.0	6.2
500	14.4	.7	6000	11.8	6.7
1000	14.2	1.2	6500	11.5	7.4
1500	13.9	1.8	7000	11.3	7.9
2000	13.7	2.31	7500	11.1	8.4
2500	13.4	3.0	8000	10.9	8.8
3000	13.2	3.5	8500	10.7	9.2
3500	12.9	4.2	9000	10.5	9.7
4000	12.7	4.6			

Loss in feet = 14.7 − Alt = X psi

$$X \times \frac{2.31 \text{ feet}}{1 \text{ psi}} = \text{Loss in feet}$$

water with a temperature in excess of 32°F gives off water vapor. Now, when the water is confined inside the pump casing, the water vapor given off causes vapor pressure. As temperature increases, so does the vapor pressure. The values of vapor pressure at various temperatures are listed in table 3.

Table 3 shows that at 212°F, the vapor pressure equals the atmospheric pressure and it will be impossible to draft water. Table 4 shows the theoretical maximum lift for varying temperatures at various altitudes. As an approximation, for each 1000-foot increase in altitude, deduct 1 foot of lift.

The temperature of the air and the level of humidity affect the performance of an internal combusion engine. Air is drawn into the cylinder and mixed with fuel; combustion, therefore, depends on the amount of oxygen available. Of prime concern is the weight of the air rather than the volume, because the increased weight means there is more oxygen. The greater the weight for a certain volume of air, the greater the density. As altitude and temperature increase, air density decreases. In addition, an increase in humidity (water vapor content) lowers the oxygen content of the air. All of these effects combine to reduce the efficiency of the engine and its ability to pump water.

The results of climatic conditions can be summarized as follows:
Pumping ability is affected by:
    1. Atmospheric pressure;
    2. Water temperature;
    3. Barometric pressure.
Engine power is affected by:
    1. Barometric pressure;
    2. Air temperature;
    3. Relative humidity.

TABLE 3. Vapor Pressure of Water at Various Temperatures

Temperature (°F)	Vapor Pressure (psi)	Vapor Pressure (feet of water)*
70	0.36	0.89
80	0.51	1.2
90	0.70	1.6
100	0.95	2.2
110	1.27	3.0
120	1.69	3.9
130	2.22	5.0
140	2.89	6.8
150	3.72	8.8
170	5.99	14.2
190	9.34	22.3
212	14.70	35.4

*The values of vapor pressure in feet of water are slightly higher than those shown in table 1 for the equivalent pressure in psi. This is due to the decrease in specific gravity of water as the temperature increases. The specific gravity at 39.2°F is 1.000 and at 212°F it is 0.958. The decrease in specific gravity means that a given volume of water weighs less, and the same atmospheric pressure will push it higher.

TABLE 4. Theoretical Lift at Varying Temperatures and Altitudes

Temperature (°F)	Lift at Altitude of −800 feet	Lift at Altitude of 0 feet	Lift at Altitude of 3400 feet
32	34.8	33.9	29.8
40	34.7	33.6	29.7
50	34.5	33.4	29.6
60	34.3	33.2	29.4
70	34.1	33.0	29.1
90	33.3	32.2	28.4
110	32.0	30.9	27.0
130	29.9	28.7	24.9
150	26.5	25.4	21.5
170	21.0	20.0	16.0
190	13.5	12.4	8.5
212	6.4	5.4	1.0

Operating procedures

To operate a centrifugal pump from draft:

1. Position the pumper as close as possible to the static source of water.
2. Connect the number of hard sleeves necessary to reach the water source (figure 3). Make sure that soft, pliable gaskets are used. Tighten the couplings to ensure an airtight seal.
3. Connect the strainer to the end of the hard sleeve (figure 4). Tighten the strainer so that an airtight seal is obtained at the coupling.
4. Tie a rope to the strainer. The rope will be used to raise the strainer off the bottom of the water source and will ease the strain on the coupling to the truck. If a floating strainer is used (figure 4), the rope is still needed to help handle the hard sleeves.

Figure 3. Drafting sleeves.

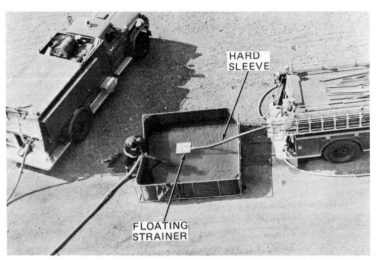

Figure 4. Drafting operation.

5. Tie a clove hitch and binder around the hard sleeve where the strainer and hose connect. Place the hitch so that it straddles the lugs of the hard sleeve. Tie the short end of the rope through the strainer eye, using a figure-8 knot. Tie another clove hitch around the female end of the first hard sleeve. The loose end of the rope can now be tied to the pumper, and the sleeve can be raised or lowered using the rope. Note: Always try to secure to the pumper in case operations must be secured and the pumper moved in emergency conditions.

6. Connect the end of the hard sleeve to the large intake of the truck. Tighten the coupling to ensure an airtight seal.

7. The strainer and hard sleeves are lowered into the water. If possible, keep the strainer at least 18 inches off the bottom and 18 inches below the surface of the water. If a floating strainer is used, it will automatically adjust to the correct angle (figure 5).

8. Close all connections and shut all drain valves to make an airtight seal.

9. Set the transfer valve to the volume position for most centrifugal pumps.

10. Engage the pump.

11. Engage and operate the priming device until the pump is primed. The vacuum reading on the intake gage should be proportional to the lift, approximately 1 inch for each foot.

12. When primed, there will be a pressure reading on the discharge gage and water will flow from the priming device. Open one discharge gate slowly, and at the same time advance the throttle until a steady flow is established. As the discharge is increased, the vacuum reading on the intake gage will increase to make up for the friction loss in the hard sleeve.

13. Set the transfer valve, the relief valve, or the governor as outlined in previous chapters.

14. To shut down the operation, lower discharge pressure slowly, close discharge gates, disengage pump, and open drains. Opening the drains will cause a loss of vacuum and drop the water from the hard sleeves.

15. Disconnect the hard sleeve from the apparatus and, using the rope, raise the strainer from the water source.

16. Return all controls and valves to their normal positions.

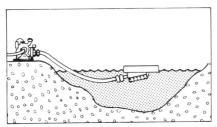

DEEP DRAFT—strainer drafts from a few inches below the surface where water is cleanest. Note how hose weight tilts strainer while float remains level.

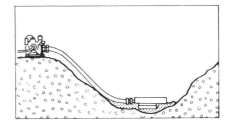

SHALLOW DRAFT—FLOAT DOCK drafts down to about five-inch depth and still no whirlpool or loss of suction.

Figure 5. Floating strainer operation.

Drafting problems

Operating a pumper from draft usually is not a frequent occurrence. For this reason, and because the pump must be in excellent condition, many operators experience trouble while trying to draft. It is recommended that frequent drills in drafting procedures, even in municipal departments, be carried out. However, if trouble should be experienced, the following trouble analysis will assist in locating the problem.

1. Air leaks — Any faulty connection or fitting can make it difficult to prime the pump. All intake hose gaskets should be checked for fit and wiped clean of sand, pebbles, and any other foreign matter before the hose is attached to the pump. The main pump packing should be adjusted tightly enough to prevent air leakage (not water leakage). A test for air leakage is outlined in Chapter 20.

2. Dirt in suction screens — Dirt-clogged suction screens may make it difficult to prime the pump as well as cause the pump to lose its prime.

3. Priming problems — Problems can be experienced when trying to prime if engine speed is too low, the primer is not operated long enough, there is no oil in the priming reservoir, there are improper clearances in the rotary primer, there is excess carbon on exhaust primer valve seats, or there is a defective priming valve.

4. High point in intake hose — High points in the hard sleeve (caused, for example, by running the hard sleeve over a railing or fence) create air pockets that can cause a loss of the prime. If the hard sleeve cannot be rearranged, prime can be obtained by closing the discharge valves immediately when the pressure drops and then repriming. This procedure usually eliminates the air pockets which were drawn into the impellers from the high point.

5. Suction lift too high for operating conditions — Too high a lift for the altitude, weather conditions, and the condition of the pump and accessories will prevent priming.

6. Strainer not under water — The strainer may be submerged deeply enough for priming, but when a large volume of water is pumped, a whirlpool is formed which uncovers the strainer. This allows air to enter the pump.

7. Suction hose collapsed — On defective or old hard sleeves, the inner liner often collapses when drafting water, thus, restricting the flow of water to the pump. Collapse of the inner liner is often hard to detect, even when the inside of the hose is carefully examined. This is due to the fact that the inner liner often goes back in place when the vacuum is removed. Refer to Chapter 20 for a method of testing the hard sleeve.

8. Pump condition — Badly worn wear rings or impeller vanes fouled with debris will cause difficulty.

9. Insufficient engine power — Engine power can be reduced due to incorrect timing, fouled spark plugs, burned points, weak condenser or coil, sticking valves, worn piston rings, worn fuel pump, or poor carburetion.

Cavitation

Cavitation is a condition that occurs internally within the pump. It is caused by the pump trying to deliver more water than is being supplied. Cavitation can occur when trying to draft more water than can be lifted for the particular circumstances, when the strainer becomes clogged, when the hard sleeve liner collapses, or when the water supply decreases.

As the flow of water to the intake decreases and the impeller rpm remains constant, pressure at the impeller eye decreases. The decreased pressure causes an increase in the amount of vacuum. Now, the increased vacuum lowers the boiling point of the water. For example, at 14.7 psi water boils at 212°F, but at 10 psi it boils at 193.22°F and at 1 psi it boils at 101.83°F.

Some of the water entering the eye of the impeller encounters the increased vacuum conditions and flashes into steam or water vapor. These bubbles of water vapor flow through the impeller where there is an almost instantaneous change from vacuum to pressure. The water vapor condenses because the increased pressure raises the boiling point. As the water rushes in to fill the void left by the water vapor, there is a tremendous shock. The shock of the water filling the empty space at a high velocity causes damage to the metal of the impeller, and the pump could be made inoperative.

Cavitation can be recognized as the point where an attempt to increase engine rpm does not produce an increase in discharge pressure. If cavitation does occur, the operator can decrease engine rpm, increase the size of the hard sleeve, locate an additional source of water, or reduce the height of the lift.

Chapter 16

Hydrant Supply and Operations

One of the greatest assets to the pump operator from a firefighting standpoint is an ample water supply piped through a distribution system to hydrants. To efficiently use the water distribution system, pump operators must have a knowledge of:
1. The operating characteristics of the hydrant,
2. The water distribution characteristics of the system,
3. The pump operating procedures at the hydrant, and
4. The care, maintenance, and testing of hydrants.

Hydrant characteristics

The design of a hydrant varies with the geographical area, the locality within the state, and even within cities and towns in the same county. Some areas have two different types of water supply systems — high pressure and regular distribution — which means that there are two different types of hydrants within the same area.

"The fire apparatus driver/operator shall identify the types of hydrants used within the jurisdiction, including descriptions of:
"(a) Connection size and type of thread of discharge openings
"(b) Construction and operation of drain valve
"(c) Direction of operation of the main valve
"(d) Internal diameter of hydrant barrel
"(e) Maximum friction loss in the hydrant
"(f) Procedures and policies of hydrant locations."*

Most hydrants are equipped with two 2½-inch discharges and one large discharge (figure 1). The large discharge is also known as the "steamer connection."

For most water systems, the threads used on the hydrant discharges are the standard 7½ threads per inch for 2½-inch openings and 4 threads to the inch for larger openings. The National Standard thread designation is NH. However, there are still many fire departments that continue to use their own nonstandard thread designations. This means that neighboring departments when called for mutual aid must have special adapters to hook up to the hydrant, which can hinder operations.

*Paragraph 3-2.2. Reprinted with permission from NFPA 1002-1982, Standard for Fire Apparatus Driver/Operator Professional Qualifications, Copyright© 1982, National Fire Protection Association, Quincy, Massachusetts 02269. This reprinted material is not the complete and official position of the NFPA on the referenced subject, which is represented only by the standard in its entirety.

Hydrants can be divided into two major categories, wet barrel and dry barrel.

Wet barrel hydrant — The wet barrel hydrant is primarily used in those areas where freezing does not occur, since water is also present in the barrel. Because freezing is not a problem, the wet barrel type has fewer parts than the dry barrel. The wet barrel hydrant has an underground section, a top section, a valve and a valve stem for each discharge, and a cap for each discharge to protect the threads and shutoff valve (figure 2).

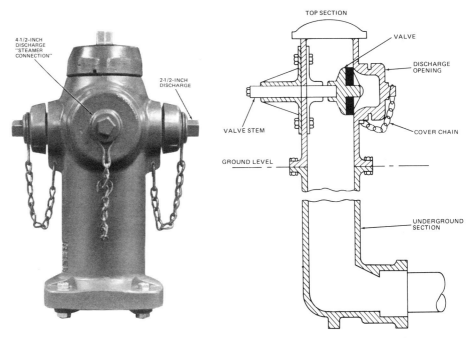

Figure 1. Modern fire hydrant. Figure 2. Wet barrel hydrant.

Dry barrel hydrant — The dry barrel hydrant consists of the foot piece, the barrel, the bonnet, the valve, the valve stem, and the discharge opening (figures 3A and 3B). The shutoff valve must be located below the frost line to prevent freezing. This necessitates additional parts, seals, and joints, making the dry barrel hydrant more complicated than the wet barrel hydrant.

The foot piece or shoe is the inlet for water from the main and contains the shutoff valve and drain valve (figure 3A). The barrel, which contains the operating stem, conducts the water from the main to the discharge opening. The bonnet provides protection to the hydrant and contains the mechanism for turning the valve. The discharge cover protects the discharge threads and keeps objects from entering and blocking the barrel.

The main valve of this particular hydrant is kept closed with water pressure, which acts on the bottom of the valve (figure 4A). This prevents water flow if the hydrant is struck and the barrel and valve stem damaged. To open the hydrant, the valve stem is turned at the operating nut, and the valve moves toward the bottom of the foot piece (figures 3A and 3B). During the first few turns, the drain valve is open and full water pressure flushes the drain waterway (figures 4A and 4B). When the main valve is fully opened, the drain valve closes (figure 4C). When operation from the hydrant is completed, the main valve can be closed, but the hydrant barrel remains filled with water. During the last few turns of the operating nut, full water pressure again flushes the drain valves (figure 4D). With the main valve fully closed, water drains from the barrel through the drain waterway, keeping the barrel dry (figure 4E).

When installing a dry barrel hydrant, it is necessary to place a coarse gravel

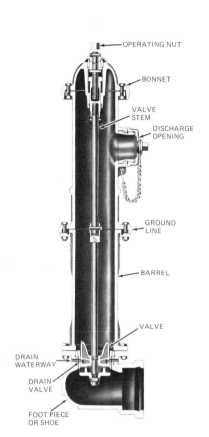
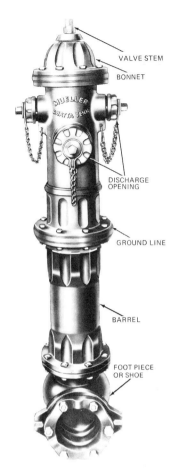

Figure 3A. Dry barrel hydrant (cross section).

Figure 3B. Dry barrel hydrant components.

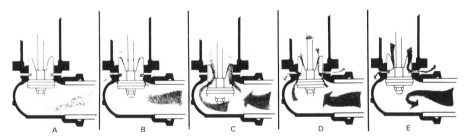

Figure 4. Hydrant main and drain valve operation.

bed around the foot piece to ensure proper draining (figure 5). The gravel also prevents the drain from becoming clogged with dirt and soil. One cautionary measure when using a dry barrel hydrant is that it must be opened more than a few turns. If it is left to run while drain valves remain open, the water leaves the waterway like a jet stream. This stream can erode the coarse rock, causing it to collapse. The breakup of the rock creates a void which can cause the street to sink or the water main to break.

In areas with a high water table, drain valves may have to be plugged. This will prevent the ground water from entering the water barrel and rising until it is level with the area water table. Water in the barrel will, of course, be subject to freezing and can also contaminate the domestic water system. When the drain holes are plugged, the fire department must hand pump the barrel dry after each use.

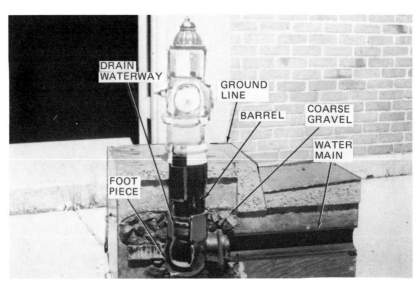

Figure 5. Hydrant installation.

Water distribution system

All of the hydrants in a city will be of little use if there is no water flowing through the mains. The water distribution system must be designed to bring sufficient water efficiently and reliably from the source to the area where needed. To do this, the system needs:

1. A source of supply,
2. A method of moving water, and
3. A distribution and storage system.

Source of supply — Two sources of supply for a water distribution system are surface waters and ground waters. Surface waters include lakes, ponds and rivers; ground water can be obtained through wells and springs.

Depending on the purity of the source, and whether the water will be used for industrial processing or human use, purification may be required.

Water movement — Water must be moved from its source to the treatment facility and from the treatment facility to the user through water mains. Two methods for accomplishing this task are to direct pump the water through the system or to pump the water to an elevated storage tank. The elevated storage tank, the most common method, then creates a pressure head equal to .434 psi for every foot of elevation.

Distribution and storage system — Distribution of water over the area is made through varying size pipes. Water is usually stored locally in elevated tanks so that a good supply, under sufficient head, is always available. The different size pipes which go together to make a water distribution system are also known as a grid (figure 6).

A grid is composed of:

1. *Primary mains* — large mains which bring water from the source or water treatment plant to the area to be served.

2. *Secondary mains* — intermediate size pipes which supply large sections of the service area.

3. *Distribution mains* — small size pipes which feed individual streets of the service area. These pipes are rarely smaller than 6 inches.

"The fire apparatus driver/operator shall identify the size of mains and the available fire flows in various areas specified by the authority having jurisdiction.

"The fire apparatus driver/operator shall identify problems related to flows from dead-end water mains.

HYDRANT SUPPLY AND OPERATIONS

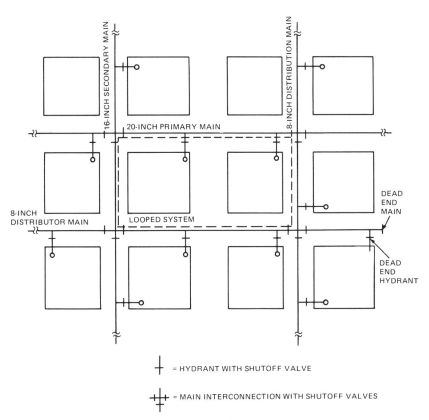

Figure 6. Water distribution system.

"The fire apparatus driver/operator, given reference material, shall identify and explain the approximate pressure-discharge relationship for various water pipe sizes.

"The fire apparatus driver/operator shall identify the pipe sizes used in water distribution systems for residential, business, and industrial districts served by the authority having jurisdiction.

"The fire apparatus driver/operator shall identify at least two causes of increased resistance or friction loss with water flowing in water mains."**

It is important to remember that one size of pipe, say 36 inches, is needed in a big city as a primary main, while a 12-inch primary main will be sufficient for a small town. Similarly, distribution mains of 6 inches may be enough for one community, while another will need 12-inch pipe. No fixed size can be established for each of the three types of mains. The size is relative to the needs of the area being served.

In order to supply sufficient water for the fire service and domestic use, water mains are *looped* (figure 6). This means that the mains are cross connected. The opposite of a looped main is a *dead-end main*, which results in lower flow (figure 6). Valves are installed throughout the system so that in the event of a break, only a small localized area must be shut down for repair.

"The fire apparatus driver/operator shall identify incrustation, tuberculation, and sedimentation, and the effects on the carrying capacities of water mains."†

**Paragraphs 3-2.3, 3-2.4, 3-2.5, 3-2.6, and 3-2.7. Reprinted with permission from NFPA 1002-1982, Standard for Fire Apparatus Driver/Operator Professional Qualifications, Copyright© 1982, National Fire Protection Association, Quincy, Massachusetts 02269. This reprinted material is not the complete and official position of the NFPA on the referenced subject, which is represented only by the standard in its entirety.

†Paragraph 3-2.1. Ibid.

The amount of water that can flow through the system is also limited by the state of the inside of the pipe. Sediment, corrosion, or rust can inhibit the flow significantly. Corrosion, which can be caused by the materials and chemicals carried in the water, is known as tuberculation. The buildup of sediment can also narrow the area of the pipe available for flow.

As explained above, a dead-end main is not the best source of water from a hydrant. Based upon the pipe size and capacity of the pumper, the maximum length of dead-end supplies to hydrants can be calculated. Table 1 gives a comparison of this information.

TABLE 1. Maximum Length of Dead-End Supply to Hydrants

Pipe Size	750 gpm Pump Feet	1000 gpm Pump Feet
4-inch	90	50
6-inch	650	380
8-inch	2,650	1,550
10-inch	7,750	4,600
12-inch	19,300	11,150

The larger the pipe, the more water that can be carried. The manual published by the International Fire Service Training Association (IFSTA 205, Third Edition), provides a comparison of the discharge capacities of pipes. Table 2 shows this comparison.

TABLE 2. Approximate Discharge Capacities of Pipes Flowing Full

Diameter in inches	Diameter in inches									
	4	6	8	10	12	16	20	24	30	36
36	—	—	—	24.6	15.6	7.6	4.3	2.8	1.6	1
30	—	—	27.2	15.6	9.9	4.8	2.8	1.7	1	—
24	—	32.0	15.6	8.9	5.7	2.8	1.6	1	—	—
20	55.9	20.3	9.9	5.7	3.6	1.7	1	—	—	—
16	32.0	11.7	5.7	3.2	2.1	1	—	—	—	—
12	15.6	5.7	2.8	1.6	1	—	—	—	—	—
10	9.9	3.6	1.7	1	—	—	—	—	—	—
8	5.7	2.1	1	—	—	—	—	—	—	—
6	2.8	1	—	—	—	—	—	—	—	—

To use the chart, find the size of one pipe in the vertical column and the size of the pipe to be compared in the horizontal column. For example, to find out how much more water a 12-inch pipe will carry than a 6-inch pipe, find the number 12 in the left column and the number 6 in the top column. Where the numbers intersect, 5.7, is the capacity difference. According to the table, the 12-inch pipe will carry 5.7 times as much water as the 6-inch pipe at the same pressure.

As another example, compare the carrying capacity of a 12-inch pipe in the left column to that of an 8-inch pipe listed in the top column. The result, 2.8, indicates that it would take more than two 8-inch pipes to carry as much as a 12-inch pipe.

Additional friction loss in pipes is created by bends, elbows and gates. Each of these fittings provides some restriction to water movement and thus increases the friction loss. A complete discussion of this friction loss is covered in Chapter 18.

High-pressure hydrants — High-pressure hydrants are generally supplied by a separate water system just for fire protection purposes. This type of system is equipped with a high-pressure pump to supply large volumes of water to high-risk areas. These systems are designed to operate from 150 to 300 psi. High-pressure hydrants usually have four individually gated discharges in addition to a main operating valve.

HYDRANT SUPPLY AND OPERATIONS

Inspection, maintenance, and testing

Responsibilities for inspecting and maintaining fire hydrants vary from community to community. Even if a fire department does not perform the inspection, they have a duty to check and ensure that hydrants will be usable when needed. A suggested procedure for performing the inspection is shown below. If a defect is uncovered, it should be noted and the proper authorities (utility company of fire department repair bureau) notified. A record of such inspection should also be kept by the inspection agency.

1. Exterior damage — Check hydrant to see if it is leaning or if there is any evidence of it being struck.

2. Ease of operation — Open and close the hydrant to check the ease with which the valve stem operates. The normal length hydrant wrench should be sufficient to open the valve. If a longer wrench is needed, the hydrant is not operating correctly.

3. Operating nut — Many different agencies make use of the water from a hydrant. Some of their personnel may use a pipe wrench to open the hydrant, thereby stripping the operating nut. The nut must be checked for this type of wear. Check also the fit of the wrench.

4. Discharge caps and threads — Many times hydrants are carelessly painted by workmen, causing the caps and chains to become frozen in place. Check each cap for ease of operation and examine the threads for nicks, breaks, and rust accumulation. Be prepared to provide a thin coat of lubrication to the threads if needed.

5. Valve operation — Check the main valve for evidence of leaking if the barrel fills with water. If flushing the valve by opening the hydrant does not help, the street valve must be closed to shut the hydrant.

6. Proper drainage — Once the hydrant has been closed, check for the opening of the drain valve by placing a hand over the 2½-inch outlet and seeing if a slight vacuum is produced. A weighted cord can also be inserted to check for complete drainage.

7. Accessibility — Check to see that the location of the hydrant can be reached with the short length of hose carried on the pumper; that the discharges are accessible (figure 7); and that there are no obstructions such as

Figure 7. Accessibility of hydrants.

fences or posts to prevent the hydrant wrench from turning the valve stem. Also see that large section may be coupled to steamer connection. Construction may have placed pavement too close to nipple to allow lugs to turn.

Fire flow tests of water systems are conducted to determine how much water is available to the fire department in particular areas. The test usually involves a series of hydrants that are flowed simultaneously. Static pressure in the main is noted and then residual pressure at a central hydrant is recorded as additional hydrants are opened.

Pressure readings are taken with a pitot gage set at the 2½-inch discharge. The pitot blade should be held at the center of the stream, about 1 inch from the discharge. The correlation of the discharge pressure with the 2½-inch opening and a correction factor of c=.90, will yield the amount of water flowing. The equation is:

$$Q = 29.83 \times d^2 \times \sqrt{P} \times c$$

Q = quantity gpm flow
d = diameter of nipple
P = pressure reading on pitot
c = correction factor

This information will provide the officer with estimates for use on the fireground and will assist him in making tactical decisions. Using the preceding formula, table 3 calculates the flow for a 2½-inch, 4-inch, and 4½-inch discharge at varying pressures.

Exact flow requirements and test procedures for evaluating a water distribution system have been developed by the American Insurance Association. The AIA performs these tests when conducting a survey of a community.

TABLE 3. Hydrant Discharges*

Pressure (psi)	2½-inch Discharge (gpm)	4-inch (Discharge (gpm)	4½-inch Discharge (gpm)
1	170	430	540
2	240	610	770
3	290	740	940
4	340	860	1090
5	380	960	1220
6	410	1050	1340
7	440	1140	1440
8	480	1220	1540
9	500	1290	1640
10	530	1360	1730
11	560	1430	1810
12	580	1490	1890
13	610	1550	1960
14	630	1610	2040
15	650	1660	2110
16	670	1720	2180
17	690	1770	2240
18	710	1820	2310
19	730	1870	2370
20	750	1920	2430
25	840	2150	2700
30	920	2350	2980
35	1000	2540	3210
40	1060	2720	3440

*The formula used was $Q = 29.83 \times d^2 \times \sqrt{P} \times c$ where c = .90. The answers were rounded to the nearest 10 gpm.

Since the location, style, position, and size of each fire hydrant can vary, the pump operator must evaluate the situation immediately from the apparatus cab. Then, he must position the apparatus so that it is in the correct position for optimum utilization of the hydrant. While operating, he must also be able to estimate the water still available from the hydrant and be on the lookout for cavitation. Each step in the operating sequence is discussed below.

Operating the pumper

1. Locate the pumper at the hydrant.
2. Select one of the three methods for connecting to the hydrant:

a. hard sleeves
b. large-diameter soft sleeve
c. 2½-inch or 3-inch soft sleeve

The use of a hard sleeve to connect to the hydrant is the most difficult method. Even if the pumper is positioned exactly, handling the hard sleeve makes the procedure very time-consuming. There is a brand of lighter weight flexible hard sleeve which overcomes some of these objections; however, both types of hard sleeve create a danger of cavitation.

The large-diameter soft sleeve method provides the most efficient way of connecting to a hydrant. A front or rear intake simplifies the connection (figure 8). Care must be taken to prevent kinking the soft sleeve and restricting the flow as well as to prevent the hose from rubbing on the ground due to truck vibration.

Figure 8. Front intake with large-diameter soft sleeve.

The 2½-inch or 3-inch soft sleeve connection eliminates the need for exact spotting. However, even if both 2½-inch hydrant discharges are used, they will not be able to flow the same amount of water that a 4½-inch outlet will flow.

3. Shift the transmission to the pump position.

4. Take off the discharge caps on the outlets of the hydrant to be used. If a large-diameter soft sleeve is used, connect a one or two-way gate valve to one of the unused 2½-inch discharges. Some departments used various types of four-way gates on the steamer connection to allow different pumpers to connect to the same hydrant. If any of the caps are frozen onto the discharge, give the caps a sharp blow with a hydrant wrench. This will break the cap and leave the threads undamaged.

5. Tighten all connections. Leaking couplings can cause hazardous conditions and result in injuries or damage to the hose.

6. Check to make sure the pumper discharge gates are closed.

7. Open the pumper intake valve and open the intake drain or pump drain to bleed off the air trapped in the hose.

8. Open the hydrant fully.

9. Note the hydrant static pressure on the intake gage.

10. Close the drain when the water flows out.

11. Open the required discharge valves.

12. Set the throttle, relief valve, governor, and transfer valve to the positions required for the particular pumping situation.

13. Note the hydrant's residual pressure with the pump flowing water. An estimate of the water still available from the hydrant can now be made. (See "Estimating available flow" below for this procedure.)

14. Maintain a watch on the various gages.

15. Shutting down is accomplished in reverse order. When putting the caps back on the hydrant discharges, tighten them just enough to discourage vandals.

Dual pumper operation

Dual pumping (or tandem pumping) is the placing of a second pumper at the hydrant and connecting both pumpers intake-to-intake (figure 9). This can be done where a hydrant is connected to a large main and a good pressure is maintained.

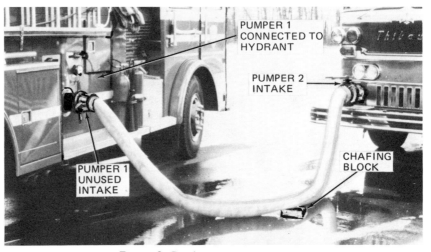

Figure 9. Dual pumper operation.

What happens when using this procedure is that the large-diameter intakes of the first pumper are used as a waterway. Some of the water that comes from the hydrant goes into the pump while the remainder flows out the other intake into the intake of the second pumper.

The advantages of a dual pumper operation are:
1. The speed with which the second pumper can get lines on the fire.
2. Less hose is needed.
3. Apparatus is grouped close together.
4. Friction loss is reduced.

The procedures for setting up a dual pumper operation are:
1. Locate the second pumper so that it can be connected to the first pumper intake-to-intake.
2. When the unused intake of the first pumper does not have a gate, the blind cap can still be removed without the first pumper shutting down. This can be accomplished by restricting the hydrant discharge or the intake pressure to about 5 psi residual. The hydrant can be slowly shut down or the intake slowly closed. When only 5 psi is indicated on the intake gage, it means that almost all the water coming in is being discharged. Now, the blind cap can be easily removed and the necessary connection made.
3. Open the hydrant or gate.
4. Follow the procedures outlined for operating a pumper.

Estimating available flow

Available flow from hydrants is estimated by determining the percent drop between static pressure and residual pressure. This information can be used to

HYDRANT SUPPLY AND OPERATIONS

determine if additional lines can be supplied or if the hydrant can be dual pumped.

The amount of water that is still available after one line is supplied is based on the following percentages:

0 to 10%	Three times the amount of water being delivered.
11 to 15%	Twice the amount of water being delivered.
16 to 25%	An equal amount to that being delivered.
over 25%	More water may be available, but not as much as is being delivered.

It is important to remember that these percentages are only guidelines. At exactly 10 percent, there is not an abrupt change from three times the water available. Also, nothing has been said about what the residual pressure would be when the maximum flow is reached. As the percentage approaches the upper limit, the residual pressure at the intake gate will be low when maximum estimated flow is reached. Again, this is just a guide to the amount of water left for firefighting operations.

The steps of the procedure are:

1. Note the static pressure on the compound gage after the hydrant is opened, but before any discharge gates have been opened.
2. Note the residual pressure on the compound gage after the first line is operating at the standard nozzle pressure.
3. Determine the percent of the drop in pressure.
4. Determine the amount of water still available.

Example: The static pressure on the compound gage when the hydrant is opened is 60 psi. When the first 2½-inch line, flowing 200 gpm, is placed in service, the residual pressure is 55 psi. Estimate the remaining gpm available.

Step 1. Determine the pressure drop from static to residual:

$$60 \text{ psi} - 55 \text{ psi} = 5 \text{ psi}$$

Step 2. Determine the percent drop:

$$\frac{5}{60} = \frac{1}{12} = 8.3\%$$

Step 3. Determine available flow:

$$0 \text{ to } 10\% = 3 \text{ times flow}$$
$$200 \text{ gpm flowing} \times 3 = 600 \text{ gpm}$$
$$600 \text{ gpm is still available.}$$

Example: The static pressure when the hydrant is opened is 90 psi. When a 500-gpm deck gun is placed in operation, the residual pressure is 72 psi. Estimate the flow still available.

Step 1. Determine the pressure drop from static to residual:

$$90 \text{ psi} - 72 \text{ psi} = 18 \text{ psi}$$

Step 2. Determine the percent drop:

$$\frac{18}{90} = 20\%$$

Step 3. Determine the available flow:

16 to 25% = same amount
500 gpm flowing × 1 = 500 gpm
500 gpm is still available

Example: The static pressure when the hydrant is opened is 90 psi. When a 500-gpm ladder pipe is placed in operation, the residual pressure is 72 psi. However, an additional pumper connects to the same system and the residual pressure drops to 45 psi. Estimate the flow still available.

Step 1. Determine the pressure drop from static to residual:

90 psi − 45 psi = 45 psi

Step 2. Determine the percent drop:

$$\frac{45}{90} = 50\%$$

Step 3. Determine the available flow:

Greater than 25% = no lines flowing 500 gpm can be added. However, some water is still available, perhaps enough for a hand line. The operator will just have to try, watching the intake gage.

Cavitation at hydrants

When pumping from a hydrant, a soft sleeve has an advantage over a hard sleeve. If a hydrant supply is poor, the pump may attempt to pump more water than the hydrant can deliver. Under these conditions, the pump will cavitate and the soft sleeve will collapse under the partial vacuum, even though the intake gage might still indicate a positive pressure. If a hard sleeve were used, the only indicator would be the intake gage, which is very inaccurate close to the zero reading. Remember, cavitation is quite common when pumping from a hydrant.

Chapter 17

Relay Operations

Relay is the movement of water from a pumper at the water source to the intake of a second pumper, out the discharge of the second pumper to the intake of a third pumper, and so on until the water reaches the fireground. Relays are necessary when the water source is too far from the fire to be moved efficiently by one pumper. The increased distance causes an increase in friction loss which one pumper by itself is unable to overcome. A high back pressure may also necessitate a relay to move the water, even if short distances are involved.

Fireground conditions usually dictate that relay operations must simply grow rather than develop according to advanced planning. This is necessary because the location, size and type of each fire cannot be planned in advance. However, with certain standard operating procedures, understanding of the relay process, and cooperation between pump operators, an involved relay operation can be successfully completed.

"The fire apparatus driver/operator, given a fire department pumper, shall demonstrate, by actual use, procedures for pumping:
  "(a) At maximum delivery rate from the apparatus water tank (see Chapter 19).
  "(b) From a hydrant, at maximum rated capacity (see Chapter 16).
  "(c) From draft, a maximum rated capacity (see Chapter 15).
  "(d) In a relay operation.
  "(e) In a tandem pumping operation (see Chapter 16).
    "1. Two pumpers in parallel
    "2. Two pumpers in series.

"The fire apparatus driver/operator, given a fire department pumper and a simulated fire scene, shall demonstrate proper maneuvering and positioning of the apparatus to function from the given source of water (see Chapters 15 through 17)."*

OPERATING REQUIREMENTS

A relay offers certain advantage to the fireground tactical situation:

1. The pumper is located at the fireground where hose lines can be easily monitored and where the proper equipment can be utilized.

2. An initial attack can be made using the water carried on the pumper until the relay is set up.

---

*Paragraphs 3-6.5 and 3-6.6. Reprinted with permission from NFPA 1002-1982, Standard for Fire Apparatus Driver/Operator Professional Qualifications, Copyright©1982, National Fire Protection Association, Quincy, Massachusetts 02269. This reprinted material is not the complete and official position of the NFPA on the referenced subject, which is represented only by the standard in its entirety.

3. Lower pump pressures can be used so that neither the engine, the pump, nor the hose lengths will be strained.

However, the time necessary to establish a relay is a vital factor, influenced by the available manpower, the hose carried, and the available apparatus. Other factors that affect the relay and thus the time required to set up include: Water needed for fire extinguishment, distance between water source and fire, hose size, pumper capability, terrain.

Quantity of water — The quantity of water needed for fire control will be determined by the officer in charge. He will determine whether hand lines or master streams will be used. The pump operator must keep the officer informed about the ability of the relay to supply the water needed.

Distance between the water source and fire — Friction loss is dependent upon the distance that water travels through the hose. Once the necessary quantity of water has been determined, the friction loss must be calculated. If the friction loss is too great with one pumper supplying the water, a relay must be set up. If the number of pumpers available is limited or if a large volume of water is needed, the relay will need two lines between pumpers.

Hose size — Friction loss can be reduced by increasing the hose diameter or by laying additional lines. The amount of hose available and its size will determine the number of pumpers necessary for the relay operation. Figure 1 shows a 1200-foot, 3-inch relay with the supply pumper at a hydrant out of sight. Note that the winding hose lay will cause an increase in friction loss.

Figure 1. Water relay.

Pumper capability — The two factors that influence the ability of a pumper to deliver water and overcome friction loss are maximum engine pressures for hose layouts and rated capacity of pumpers.

The limitation of engine pressures due to hose capacity was discussed in Chapter 6. A maximum of 200-psi working pressure should be established, based on an annual hose test of 250 psi, leaving a safety factor of 50 psi. When pumping in relay, 20-psi residual pressure is needed to supply the next pumper in a relay or to supply a reservoir. Therefore, the 20-psi residual pressure can be considered an intake or reserve pressure. It is not good practice to supply the next pumper with just enough pressure because if the pattern on the nozzle changes, causing a need for slightly more water, the pumper without the reserve pressure would run away from the water. This, in turn, will cause cavitation.

The balance of the available engine pressure (180 psi) can be used to overcome friction loss. Table 1 shows the distances the 180 psi will deliver water through the various size hoses.

TABLE 1. Maximum Distance Water Delivery

Flow*	Hose Size		
	2½-inch	3-inch	4-inch
gpm	feet	feet	feet
100	6000	18,000	— —
200	1800	4500	18,000
250	1200	3000	12,000
300	850	2000	9000
400	500	1100	4500
500	300	720	3000
750	150	350	1500

*If two lines are laid, divide the total flow in half and then read the maximum value for each line.

The rating of a pumper is based on the amount of water that it can pump. As the volume increases, the pressure that the pump can develop decreases. Therefore, the capacity of the pumpers used in a relay reaches a critical value as the flow requirements increase. Pumpers are rated at

100 percent of maximum capacity at 150 psi
70 percent of maximum capacity at 200 psi
50 percent of maximum capacity at 250 psi

For example, if a flow of 1000 gpm were needed on the fireground, a 1000-gpm pumper would only be able to develop about 150-psi discharge pressure (depending on whether the source is static or pressure). This means that large lines between pumpers and close spacing would be necessary. The smallest pump that can deliver 1000 gpm at 200 psi is a 1500-gpm pump.

For this reason, the largest pumper should be placed at the water source. The smaller capacity pumpers should be used to supply lines at the fireground. This will make maximum use of all pumpers.

Using these principles, the officer responsible for establishing the relay can estimate the number of pumpers needed.

Example: A 500-gpm master stream device is needed 2000 feet from a water source. How many pumpers would be needed (a) if a single 2½-inch line were used? (b) if two 2½-inch lines were used? (c) if a 4-inch line were used?

Step 1. Using table 1, determine the maximum distance for the hose size and flow:

(a) one 2½-inch = 500 gpm = 300 feet
(b) two 2½-inch = 250 gpm = 1200 feet
(c) one 4-inch = 500 gpm = 3000 feet

Step 2. Divide the water to fire distance by the water delivery distance. This yields the number of water supply pumpers.

(a) 2000 ft ÷ 300 ft = 7 pumpers
(b) 2000 ft ÷ 1200 ft = 2 pumpers
(c) 2000 ft ÷ 3000 ft = 1 pumper

Step 3. Add one pumper for the attack pumper on the fireground. This yields the total number of pumpers needed for the relay. In some cases this extra pumper will not be necessary, but again, it should be included as a safety factor.

(a) 7 + 1 = 8 pumpers
(b) 2 + 1 = 3 pumpers
(c) 1 + 1 = 2 pumpers

From the preceding example, it is clear that large-diameter hose or dual relay lines can reduce the number of pumpers needed to move water. Figure 2 shows a specialized relay hose wagon using 4-inch hose on a reel. Of course, if in the above examples only 500-gpm pumpers were available, more pumps would be needed because the pump can deliver only its rated capacity of 500 gpm at 150 psi. One other limitation is the amount of hose carried on each pumper.

Figure 2. Large-diameter hose supply for a relay.

Terrain — The rise and fall of the ground between the water source and the fireground also affects the number of pumpers needed. However, table 1 has enough of a margin to account for some added pressure needed to move water uphill. Large, hilly areas may require an additional pumper.

OPERATING PROCEDURE

When the hose is laid and the pumpers are in position, the following procedure for establishing the relay can be followed:

1. Open two discharge gates on all pumps, except on the pump at the source, to get rid of air from hose lines and pumps. On each pumper, attach the hose line to one of the discharges and leave the other discharge uncapped. Uncapping the second discharge gate is not necessary if a relay valve (figure 3) is installed. The relay valve is a relief valve connected on the intake side of the

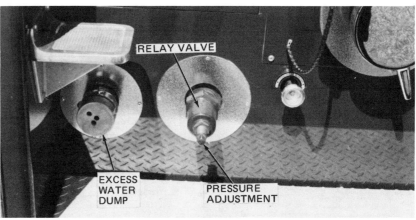

Figure 3. Relay valve.

pump. It automatically opens and dumps water on the ground if too high a pressure is supplied, thus protecting the pump. If no relay valve is present, the operator can watch his intake gage and if a high pressure is reached, the gate controlling the uncapped discharge can be opened and the excess water dumped on the ground.

2. The pump operator at the water source gets water to his pump and then discharges it so that water starts moving toward the fireground. Just to start the water moving, discharge pressure should not be over 150 psi.

3. As soon as water reaches the second pump, the operator closes the uncapped discharge gate. Water is now being discharged to the next pumper. The throttle should be advanced until the 150-psi discharge pressure is obtained.

4. Each pump operator in turn duplicates step 3 until the water is delivered at the fireground.

5. The pump operator at the fire scene then advises all other pump operators of the amount of water needed at the fireground.

6. The pump operator at the water source now adjusts the throttle until the correct operating pressures for the current situation are obtained. Remember, the discharge pressure should not exceed 200 psi, if possible. Care must be exercised that this pump does not run away from its source, causing cavitation. If more water is needed than the relay can supply, advise the officer so that arrangements for an additional supply can be made.

7. Each subsequent pump operator then adjusts his pressure to meet the particular situation, without going below 20-psi residual pressure on the intake gage. The pump operator can check the hose connected to the pump intake by feel, to determine when maximum delivery is reached.

8. The pump operator at the fire adjusts the discharge pressure to supply the lines being used.

9. The operation of adjusting the pressure is repeated as often as necessary. Gages must be observed carefully during a relay. The relief valve or governor should be set when pumping.

10. Once water is moving, every effort should be made to keep it moving throughout the relay operation. Nozzles should not be shut off unless absolutely necessary. For a temporary shutdown, the operator of the fireground attack pumper can let the excess water dump on the ground. If a hose line bursts, the operator of the last pump should open a discharge gate to waste water while the hose section is replaced.

11. Shutting down is done by working from the fireground pump to the source. Pressure on the fireground pump is reduced gradually until the pump can be disengaged.

12. Each pump operator, in turn, reduces the pressure gradually, disconnects the pump, and opens an unused discharge gate until the pump source is reached.

RELAY HYDRAULICS

The hydraulics for relay pumping are the same for any other water movement. The formula for the engine pressure of the relay pumper can be expressed as

$$EP = FL + RP \pm E, \text{ where}$$
$$EP = \text{engine pressure in psi}$$
$$FL = \text{friction loss}$$
$$RP = \text{residual pressure (not less than 20 psi)}$$
$$\pm E = \text{elevation (gain or loss)}$$

PUMP OPERATORS HANDBOOK

The attack pumper at the fire scene will calculate the engine pressure using the standard formula

$$EP = FL = RP \pm E$$

Example: In the layout below, what is the engine pressure necessary at each of the pumpers?

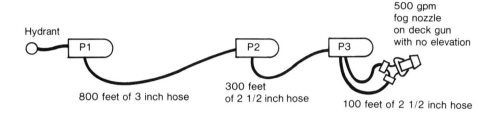

Step 1. Select the proper equation:

$$EP = FL + RP \pm E \text{ (for pumpers 1 and 2)}$$
$$EP = FL + RP \pm E \text{ (for pumper 3)}$$

Step 2. Determine the formula values for pumper 3:

$$\begin{aligned} RP &= 100 \\ E &= 0 \\ FL &= 10 \text{ psi for device} \\ &\ \underline{15} \text{ psi for } 2\tfrac{1}{2}\text{-inch hose per 100 ft} \\ &= 25 \text{ psi} \end{aligned}$$

Step 3. Solve the formula for engine pressure for pumper 3:

$$EP_3 = 25 + 100 + 0 = 125 \text{ psi}$$

Step 4. Determine the formula values for pumper 2:

$$\begin{aligned} RP &= 20 \text{ psi} \\ E &= 0 \\ FL &= 55 \text{ psi}/100 \text{ ft} \times 3 = 165 \text{ psi} \end{aligned}$$

Step 5. Solve the fomrula for engine pressure for pumper 2:

$$EP_2 = 165 + 20 + 0 = 185 \text{ psi}$$

Step 6. Determine the formula values for pumper 1:

$$\begin{aligned} RP &= 20 \text{ psi} \\ E &= 0 \\ FL &= 25 \text{ psi}/100 \text{ ft} \times 8 = 200 \text{ psi} \end{aligned}$$

Step 7. Solve the formula for engine pressure for pumper 1:

$$EP_1 = 200 + 20 + 0 = 220 \text{ psi.}$$

Example: In the layout below, what is the engine pressure necessary at each of the pumpers?

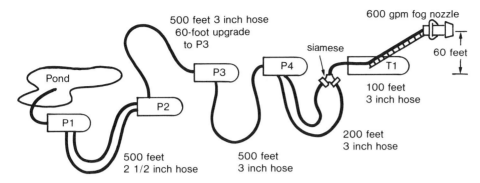

Step 1.

$$EP_4 = FL + RP \pm E$$
$$EP_{1,2,3} = FL + RP \pm E$$

Step 2.

$RP_4 = 100$ psi
$E = +30$ psi
$FL_4 = 10$ psi for ladder pipe
$\ 30$ psi for 3-inch hose
$\ \ 5$ psi for siamese
$\ \underline{\ 9}$ psi for dual 3-inch lines
$= 54$ psi

Step 3.

$$EP_4 = 54 \text{ psi} + 100 \text{ psi} + 30 \text{ psi} = 184 \text{ psi}$$

Step 4.

$RP_3 = 20$ psi
$E = 0$
$FL = 30$ psi/100 ft $\times$ 5 = 150 psi

Step 5.

$$EP_3 = 150 + 20 + 0 = 170 \text{ psi}$$

Step 6.

$RP_2 = 20$ psi
$E = +30$ psi for 60-foot upgrade
$FL = 30$ psi/100 ft $\times$ 5 = 150 psi

Step 7.

$$EP_2 + 150 + 20 + 30 = 200 \text{ psi}$$

Step 8.

$RP_1 = 20$ psi
$E = 0$
$FL = 21$ psi/1000 ft $\times$ 5 = 105 psi

Step 9.

$$EP_1 = 105 + 20 + 0 = 125 \text{ psi}$$

Example: In the layout below, what is the engine pressure necessary at each of the pumpers?

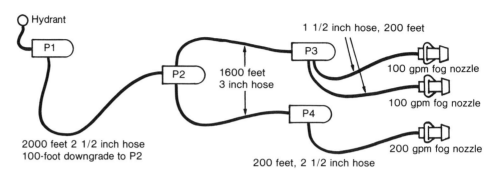

Step 1.

$$EP_{3,4} = FL + RP \pm E$$
$$EP_{1,2} = FL + RP \pm E$$

Step 2.

$$RP_4 = 100 \text{ psi}$$
$$E = 0$$
$$FL_4 = 10 \text{ psi}/100 \text{ ft} \times 2 = 20 \text{ psi}$$

Step 3.

$$EP_4 = 20 + 100 + 0 = 120 \text{ psi}$$

Step 4.

$$RP_3 = 100 \text{ psi}$$
$$E = 0$$
$$FL_3 = 30 \text{ psi}/100 \text{ ft} \times 2 = 60 \text{ psi}$$

Step 5.

$$EP_3 = 60 + 100 + 0 = 160 \text{ psi}$$

Step 6.

$$RP_2 = 20 \text{ psi}$$
$$E = 0$$
$$FL_2 = 4 \text{ psi}/100 \text{ ft} \times 16 = 64 \text{ psi}$$

Step 7.

$$EP_2 = 64 + 20 + 0 = 84 \text{ psi}$$

Step 8.

$$RP_1 = 20 \text{ psi}$$
$$E = -50 \text{ psi for 100 ft downgrade}$$
$$FL_1 = 10 \text{ psi}/100 \text{ ft} \times 20 = 200 \text{ psi}$$

Step 9.

$$EP_1 = 200 + 20 - 50 = 170 \text{ psi}$$

Chapter 18

Sprinkler and Standpipe Operations

Automatic sprinkler and standpipe systems provide firefighters with assistance in controlling and extinguishing fires. A basic knowledge of how these systems operate and the methods necessary for supplying water is essential for the pump operator.

"The fire apparatus driver/operator, given a check valve on the fire department connection to an automatic sprinkler system, shall demonstrate the direction of flow of water through the valve.

"The fire apparatus driver/operator shall demonstrate the method specified by the authority having jurisdiction for augmenting water supplies to sprinkler systems.

"The fire apparatus driver/operator, given specific information on a sprinkler system, shall identify the number of sprinkler heads that can be adequately supplied with water by various capacity rated fire department pumpers.

"The fire apparatus driver/operator, given specific information on a sprinkler system, shall demonstrate the minimum hose layouts and pump discharge pressure required to adequately supply that sprinkler system.

"The fire apparatus driver/operator shall demonstrate the method specified by the authority having jurisdiction for supplying water to a dry standpipe system.

"The fire apparatus driver/operator shall demonstrate the method specified by the authority having jurisdiction for supplementing water supplies to a standpipe system."*

STANDPIPES

Standpipe systems offer an immediate water source. They usually are installed in tall buildings (the exact height depends on local building codes), buildings with a large floor area, and places of public assembly. The standpipe can be thought of as a water main within the building for fire department operations.

The advantages of a standpipe system are:

1. Water can be quickly applied to the fire by civilians or fire brigade personnel before the arrival of the fire department.

---

*Paragraphs 3-3.1, 3-3.2, 3-3.3, 3-3.4, 3-3.5, and 3-3.6. Reprinted with permission from NFPA 1002-1982, Standard for Fire Apparatus Driver/Operator Professional Qualifications, Copyright©1982, National Fire Protection Association, Quincy, Massachusetts 02269. This reprinted material is not the complete and official position of the NFPA on the referenced subject, which is represented only by the standard in its entirety.

2. Time can be saved because firefighters do not have to stretch lines between floors of the fire building.

3. The large diameter and straight path of the standpipe reduces friction loss.

Types of standpipes

The type of standpipe installed in a building depends on its intended use. The National Fire Protection Pamphlet 14, "Standpipes and Hose Systems," establishes three classes of standpipe systems which are summarized in table 1.

TABLE 1. Standpipe Classification

Type	Use	Hose Size	Pipe Size	Water Supply
Class I	Used for heavy streams by trained fire department personnel for large volumes of fire.	Has 2½-inch discharge and all portions of the floor must be within 30 feet of the nozzle with 100 feet of hose.	For buildings up to 100 feet, 4-inch pipe minimum must be used. Over 100 feet, 6-inch pipe minimum is necessary.	If the building contains a single standpipe, a supply of 500 gpm is necessary. Each additional standpipe requires 250 gpm. When the required water is flowing, there must be 65 psi residual pressure at the highest discharge outlet. The total required flow must last 30 minutes.
Class II	Used for small streams by the building occupants for small fires.	Has 1½-inch discharge and all portions of the floor must be within 20 feet of the nozzle with 75 feet of hose.	For buildings up to 50 feet, 2-inch pipe minimum must be used. Over 50 feet, 2½-inch pipe minimum is necessary.	A flow of 100 gpm is necessary with a 65 psi residual pressure at the highest discharge outlet during maxmum flow. The total flow must be available for 30 minutes.
Class III	Used for both heavy and small streams.	Has a 2½-inch discharge with a 2½-inch to 1½-inch reducer for smaller hose. All portions of the floor must be within 30 feet of the nozzle, with 100 feet of hose.	Same as Class I	Same as Class I

In addition to the various classes of standpipes outlined in the table, there is a variety of methods for supplying them:

1. Water under pressure in the standpipe system at all times.

2. Water admitted to the standpipe system automatically as soon as a discharge valve is opened.

3. Water admitted to the standpipe system through the manual operation of a remote control device.

4. Standpipe is dry at all times. Water must be supplied by a fire department pumper.

SPRINKLER AND STANDPIPE OPERATIONS

Components

A typical standpipe installation for a high-rise building with two stairwells is shown in figure 1. The major components of this system are:

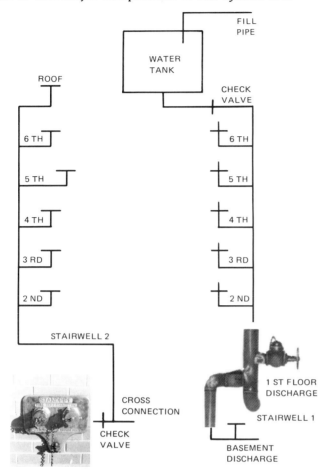

Figure 1. Double-riser standpipe system.

Siamese connection — The siamese connection is the exterior connection for the standpipe system that permits the fire department to pump water into the system (figure 2). While the exterior shape can vary, depending upon location, each siamese has two 2½-inch connections with clapper valves so that the system can be charged with only one line connected. Then, a second line can be added to augment supply without interrupting the original flow.

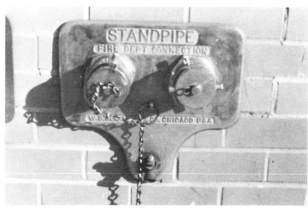

Figure 2. Standpipe siamese connections.

211

Check valve — The check valve prevents water from flowing in the wrong direction. The check valve at the siamese is kept closed by back pressure exerted by the elevated water in the system. Pressure from a pump into the siamese forces this valve open when the incoming pressure exceeds the back pressure. The check valve at the roof water system is normally open to allow water into the standpipe system. Pressure from a pump closes this check valve to prevent overfilling the storage tank and causing damage. A drip valve is also installed between the check valve and siamese to prevent water from accumulating in the line if a leak develops.

Riser — The riser brings the water from the siamese connection to the various floors of the building (figure 3). The size of the riser pipe is determined by the class of the system (table 1).

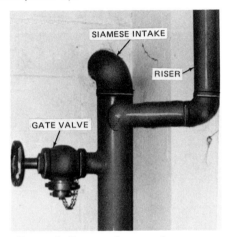

Figure 3. Riser and gate valve.

Pressure reducer — The pressure reducer prevents excessive pressure from being supplied to a hose stream by inexperienced civilians. Fire department personnel should remove the device before connecting hose lines and control flow by opening and closing the globe valve.

Hose — Some standpipe systems have 1½-inch hose (known as house hose) already connected to the globe valve (figure 4). Since this hose is made

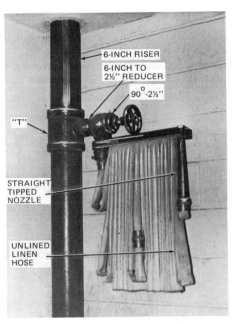

Figure 4. Preconnected hose lines.

SPRINKLER AND STANDPIPE OPERATIONS

of unlined linen, it cannot withstand normal fire department pressures, and usually has a straight tip nozzle of a very small diameter. The hose should be removed and replaced with fire department hose before beginning the fire attack. Again, because water accumulation in the linen hose will cause deteriorating conditions, a drip valve is installed between the globe valve and the hose. The drip valve must be closed before opening the globe valve.

Roof outlet — Some standpipe systems are extended to the roof (sometimes known as a header). Lines from the roof connection can be used for roof fires and to protect adjoining exposures. To prevent freezing in the exposed portion of the pipe, it is kept dry, with water flow controlled by a valve from the roof.

Water supply

The water supply for a standpipe system varies with each installation. Possibilities include:

Domestic water system — If adequate residual pressure can be maintained when the standpipe is flowing the required amount of water, the domestic water system can be used.

Dry pipe system — The fire department pumper will be the source of water for this system.

Fire pumps — When fire pumps are used to supply the standpipe system, they are usually located below ground level. The pump must be capable of delivering the required flow with the required residual pressure at the highest discharge outlet, depending on the class of the installation., There are three ways that the fire pump can be supplied with water:

1. Fed by two different mains, with each main being supplied from two directions.

2. Fed by one main and a storage tank with a capacity of 30 minutes for operating at the rated capacity of the pump. In an emergency, there must be a method of bypassing the storage tank and connecting the pump directly to the main.

3. Fed by one main, a storage tank with a capacity of 15 minutes, and a fire department siamese connection. Again, there must be a method of bypassing the storage tank and connecting the pump directly to the main in an emergency.

Elevated tanks — Located on the roof of a building, the tank provides pressure due to elevation of the water. This pressure is .434 psi for each foot above the discharge. Therefore, to get adequate pressure at the uppermost discharge, the tank must be elevated above the roof. Water is supplied to the tank from a fill pump. The fill pump usually has a small capacity, so that more water will be discharged than the fill pump can supply. This means that the tank will empty and a fire department pumper must connect to the siamese connection to continue to supply the system. Freezing is a problem if the tank is only used for standpipe supply. Since there will be long periods of time without water flow, the tank must be kept in a heated enclosure or a heated device inserted inside the tank.

Pressure tanks — Located on the roof or on the top floor of the building, these tanks are filled with water and air under pressure. When a discharge is opened, the compressed air forces the water out of the tank under pressure. In addition, there is pressure due to the tank elevation. As with the elevated tank, the pressure tank must be kept from freezing. Water is supplied to the system by a fill pump and air pressure by a compressor.

Fireground operating procedures

1. Pre-fire planning and inspection is necessary to identify and locate all

standpipes in the area served, as well as determine the location of hydrants and other water supplies. A good fireground operation starts with pre-fire planning.

2. One of the first-operating pumpers should stretch a line from the standpipe siamese to the water source. (Note: If the standpipe is to be used, the first line must be stretched to the siamese.) A second line should be laid if the officer feels the flow will be high.

3. Remove the cap on the siamese intake. Some caps are removed with spanner wrenches, while others require that the cap be hit with a heavy object and broken. Check the female swivel on the intake to make sure that there is a gasket and it rotates smoothly.

4. Connect lines from the pumper to the siamese. Charge lines only when ordered to do so by an officer.

5. Charge the standpipe system (see Engine pressures for standpipes below).

6. If the siamese connection is inoperable, the standpipe supply line can be connected to the first floor discharge using a double female fitting attached to the globe valve. Tie one end of a rope around the double female fitting and the other end around the riser to relieve the strain on the globe valve fitting.

7. Hose for use in the fire attack is carried to the floor below the fire if the standpipe system is exposed. If the standpipe system is enclosed in a fire-resistive stairway, the elevator can be taken to the floor below the fire and then the hose carried up the stairway and connected on the fire floor. It will also be necessary to carry spanner wrenches, nozzles, and various fittings, depending upon the installations in particular localities. Many departments have developed various devices to carry this equipment (figure 5).

8. A firefighter should regulate the flow at the globe valve being used to prevent accidental shutdown of the attack line.

9. After the fire, shut the globe valve, remove fire department hose and fittings, and replace with the house hose lines.

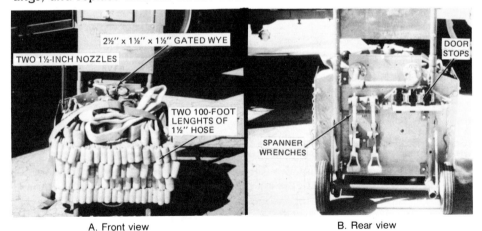

Figure 5. Standpipe cart.

Engine pressures for standpipes

Since complicated calculations cannot be carried out on the fireground, basic rules of thumb have been developed for pumping into a standpipe. These rules are stated in table 2.

To the pressure at the standpipe siamese must be added the friction loss for the supply lines from the pumper to the siamese connection. If it is necessary to actually calculate the friction loss, an allowance for the standpipe of 15 psi can be used. While this figure will increase for large flows, this is an accurate estimate for most flows that departments will handle.

SPRINKLER AND STANDPIPE OPERATIONS

TABLE 2. Rule-of-thumb engine pressure for standpipe operation

Nozzle Type	Fire Floor	Pressure Needed at Standpipe Siamese
Fog nozzle 100 psi NP	First through tenth	150 psi
Fog nozzle 100 psi NP	Above tenth	200 psi
Straight tip 50 psi NP	First through tenth	100 psi
Straight tip 50 psi NP	Above tenth	150 psi

Example: What engine pressure is required to deliver 200 gpm to a 200-foot length of 2½-inch hose with a fog nozzle that is stretched from the sixth floor discharge (five floors above the ground)? The pumper is connected to the standpipe siamese with 300 feet of 2½-inch hose.

Method 1

Step 1. Select the correction equation and determine the needed pressure by using table 2.

$$EP = FL + 150$$

Step 2. Determine the formula values

$$FL = 10 \text{ psi}/100 \text{ ft} \times 3$$
$$= 30 \text{ psi from pumper to siamese}$$

Step 3. Solve the equation:

$$EP = 30 + 150 = 180 \text{ psi}$$

Method 2

Step 1. Select the correct equation

$$EP = FL + NP \pm E$$

Step 2. Determine the formula values:

$$NP = 100 \text{ psi}$$
$$E = +5 \text{ psi/floor} \times 5 \text{ floors}$$
$$= 25 \text{ psi}$$
$$FL_{2\frac{1}{2}} = 10 \text{ psi}/100 \text{ ft} \times 2$$
$$= 20 \text{ psi}$$
$$FL_{stp} = 15 \text{ psi}$$
$$FL_{2\frac{1}{2}} = 10 \text{ psi}/100 \text{ ft} \times 3$$
$$= 30 \text{ psi}$$
$$FL_{total} = 20 + 15 + 30 = 65 \text{ psi}$$

Step 3. Solve the equation:

$$EP = 65 + 100 + 25 = 190 \text{ psi}$$

While method 2 will give a more accurate figure, method 1 provides the pump operator with a very quick calculation for immediate use. As the height the water is being pumped changes, the accuracy of method 1 changes, but it can always be used as a reliable estimator.

Friction loss in the standpipe itself can be accurately calculated for flow tests. Table 3 lists the friction loss for steel pipe, and table 4 shows the equivalent length of various pipe fittings. To solve the problem, the fittings on the standpipe system are noted and then they are changed to an equivalent length of pipe. This length is then added to the total pipe length. Once the total pipe length is known, the friction loss can be determined.

TABLE 3. Friction Loss in Steel Pipe

GPM	2 inches (per foot)	2½ inches (per foot)	4 inches (per foot)	6 inches (per foot)
50	.047	.019	—	—
100	.174	.071	.006	—
150	.380	.155	.013	—
200	.663	.267	.023	.003
250	1.03	.413	.030	.004
300	—	.585	.049	.006
350	—	.792	.066	.009
400	—	1.03	.085	.011
450	—	—	1.07	.014
500	—	—	1.30	.017
550	—	—	1.57	.020
600	—	—	1.86	.023
700	—	—	2.50	.031
800	—	—	3.24	.040
900	—	—	4.08	.051
1000	—	—	5.02	.062

TABLE 4. Pipe fitting equivalent lengths

Fitting	2½" Equiv. Feet	4" Equiv. Feet	6" Equiv. Feet
90°	9.3	13.0	17.0
45°	3.2	5.5	8.7
"T"	13.0	21.0	31.5
Check Valve	22.0	38.0	63.0
Globe Valve	62.0	110.0	—
Gate Valve	1.7	2.5	—

Example: Using the preceding example, what is the friction loss if the standpipe is 4 inches, the flow 200 gpm, and the discharge five floors above the ground (50 feet of pipe)?

Step 1. Prepare a schematic diagram of the standpipe installation:

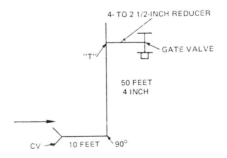

Step 2. Calculate the total feet for the 4-inch pipe:

CV (check valve) = 38 ft (table 4)
Horizontal = 10 ft
90° = 13 ft (table 4)
Vertical = 50 ft
T = 21 ft (table 4)
Total = 132 ft

Step 3. Calculate the friction loss in the 4-inch pipe:

$$FL = .023/ft \times 132 \text{ ft (table 3)}$$
$$FL = 3.036 \text{ psi}$$

Step 4. Calculate the total feet for the 2½-inch pipe:

$$90° = 9.3 \text{ ft (table 4)}$$
$$\text{Gate valve} = 1.7 \text{ ft (table 4)}$$
$$9.3 + 1.7 = 11$$

Step 5. Calculate the friction loss in the 2½-inch fitting:

$$FL = .267/ft \times 11 \text{ ft (table 3)}$$
$$FL = 2.937 \text{ psi}$$

Step 6. Add the friction losses together for the total loss in the standpipe:

$$FL_4 + FL_{2½} = 3.036 + 2.937 \text{ psi}$$
$$FL_{total} = 5.973 \text{ psi}$$

SPRINKLERS

When installed, a sprinkler system is the first line of defense for fire extinguishment. The system provides a means for immediately supplying water to the fire and, as such, it can be likened to having a firefighter on the job with a charged line 24 hours a day.

Types of sprinkler systems

The type of sprinkler system installed in a building depends on the material to be protected, the type of structure, and the availability of the water supply. Sprinkler systems can be classified as follows:

1. *Wet pipe systems* are where the piping remains filled with water under pressure at all times. Water is discharged immediately. The building must be kept heated to prevent the water in the pipes from freezing.

2. *Dry pipe systems* have their pipes filled with air, either compressed or atmospheric, which keeps a valve closed. Opening a sprinkler head allows the air pressure to drop, permitting the valve to open and let water into the piping. Water is not discharged immediately, but the problem of freezing pipes is controlled.

3. *A deluge system* distributes water to a large area all at once. This system differs from a regular system that supplies water in steps as the fire progresses and causes sprinkler heads to open. A deluge system is used in locations where a large fire can develop quickly, such as explosives manufacturing.

4. *Preaction systems* are dry pipe systems provided with a rate-of-rise indicator. At a certain predetermined rate of temperature rise, the water valve is opened, the system is charged, and an audible warning alarm is transmitted. Water then charges the system to all the heads before the one in the fire area actually fuses in most instances. This system is used in those installations where water can cause damage if the sprinkler system were operated by accident or before it was really needed.

5. A *rate-of-rise system* uses a copper ball that expands a present amount as the room temperature increases over a certain level. This expansion can be used to activate various fire safety devices; it can close fire doors, open ventilators, or open the valve to let water into the sprinkler system.

6. *Special foam systems* are installed where hazardous materials are

handled. These systems are normally kept dry, with water being admitted to the system by manual control.

Components

Control valve — The control valve regulates flow of water from the water main to the sprinkler system. The two main types of valves are the outside stem and yoke valve and the post indicator valve (figures 6 and 7). The OS&Y valve indicates that it is open when the stem sticks out of the hand wheel. The PI valve indicates that it is open when the word "open" shows in the window.

Figure 6. OS&Y valve in open position.   Figure 7. PI valve in open position.

Siamese connection — The siamese connection is the exterior connection to the sprinkler that permits the fire department to pump water into the system (figure 8). When a building has both a sprinkler and standpipe, the operator must be careful to connect the correct system.

Figure 8. Sprinkler siamese connection.

Dry pipe valve — The dry pipe valve uses a difference in surface area to keep the valve closed. A smaller air pressure, acting on a larger area (figure 9A) overcomes the water pressure acting on a smaller area. This valve keeps the water out of the system. When a sprinkler head operates, the air pressure in the piping above the dry valve is reduced. The pressure of the water supply raises and rotates the clapper and water flows to the sprinklers (figure 9B). A dry pipe valve is shown in figure 9C.

Accelerators and exhausters — One of the major drawbacks of a dry pipe system is the time it takes to bleed off the air, open the dry pipe valve, and have water flowing out the sprinkler head. To overcome this problem, an ac-

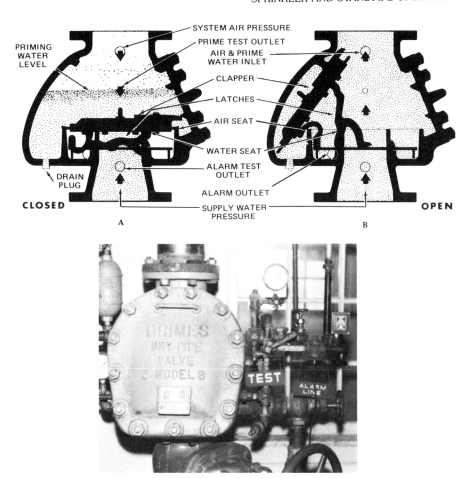

Figure 9. Dry pipe valve.

celerator or an exhauster in installed on the system to open the dry pipe valve when very small changes in air pressure occur (figures 10A and 10B). This, then, allows the water to flow to the heads much quicker.

The bottom chamber of the accelerator is connected to the sprinkler system. The air pressure in the bottom, middle, and top chamber is equalized through openings A and B in figure 10C). When a sprinkler opens, the air pressure in the system is reduced, with a corresponding reduction of pressure in the middle and bottom chambers. The pressure in the top chamber does not reduce as quickly as in the other chambers because of the restriction (figures 10C and 10D). This forces down the main valve assembly and opens the auxiliary valve. The auxiliary valve allows air in the middle chamber to exhaust rapidly into the outlet chamber, accelerating the downward movement of the main valve assembly. As the main valve opens, the air in the sprinkler system passes through the bottom and outlet chambers of the accelerator to the intermediate chamber of the dry pipe valve and equalizes the air pressure on the bottom and top of the clapper (figure 9A). This immediately destroys the pressure differential and reduces the operating time.

Water flow alarm — Both dry and wet systems fed by a variable pressure water supply contain an audible flow alarm on the exterior of the building to indicate water flow in the sprinkler system (figure 11). If the building is unoccupied, the people in the street can notify the fire department that the system is activated. Emergency personnel may also use the exterior indication of water from the drain on the water motor gong to indicate if there is indeed water movement in the system. One drawback is obvious. Fire department notification is dependent upon passers-by. However, most sprinkler installations have

PUMP OPERATORS HANDBOOK

A. Exhauster

Accelerator

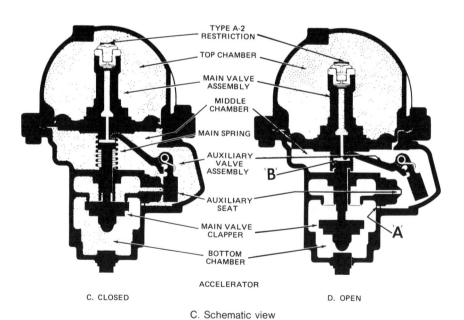
C. Schematic view

Figure 10. Rapid air-removal devices.

an additional internal alarm signal that is transmitted to a security company and then to the fire department. This is known as a supervised alarm system.

SPRINKLER AND STANDPIPE OPERATIONS

Figure 11. Water flow alarm.

Retard chamber — This is located on variable pressure supplied systems only. It is installed between the sprinkler valve and the alarm activation mechanism. It is designed to prevent accidental transmission of alarms due to surges in the variable pressure water supply. Minor surges not equivalent to the discharge from a ½-inch head, enters the valve and is diverted to a chamber prior to completing the circuit to the alarm system. A time delay set from 15 to 60 seconds depending on the size of the chamber retards the transmission of alarm until such surges can stabilize. However, in no case can the retard chamber hold back the alarm from being transmitted due to the flow of water approximating that of a ½-inch head (15 to 22 gpm).

Risers — The risers carry the water from the intake valve and distribute it throughout the building. The size depends on the number of heads to be supplied.

Sprinkler heads — Sprinkler heads are devices that are sensitive to heat. When the temperature reaches a certain predetermined value, the sprinkler head opens and allows water to flow. The most common type of sprinkler head is the fusible link (figure 12). This type of head operates as follows:

1. When heat increases to the fusing temperature, the special temperature

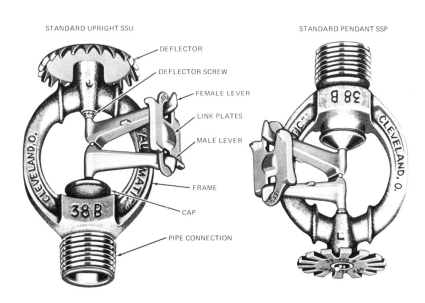

Figure 12. Fusible link sprinkler heads.

rated solder melts. The link plates, through the force imposed by the levers separate (figure 13A).

2. Fixed tension of the frame, acting as a spring, ejects levers and links clear of the sprinkler (figure 13B).

3. Water under pressure forces the disk off the orifice seat (figure 13C). Water is allowed to flow across the head and strike the diffuser.

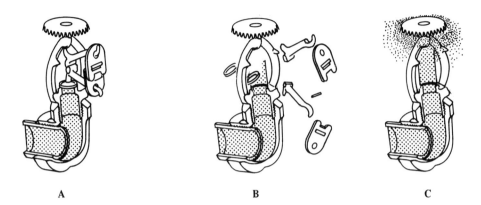

Figure 13. Operation of a sprinkler head.

Fusible link type sprinkler heads are designed to operate either upright if there is enough clearance, or suspended from the water pipe. The suspended head is called a pendant type. Figure 12 shows each of these types. Note that the deflector on each is different, so that when replacing a head, the correct type must be used. If in doubt, the deflector on the upright will be marked "SSU" while on the pendant it will be marked "SSP." The water pattern for each of these sprinkler heads is shown in figure 14.

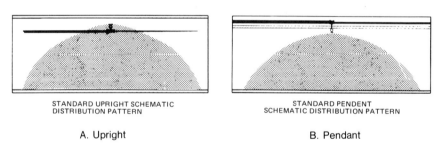

Figure 14. Water distribution from a sprinkler head.

Another type of sprinkler head is the frangible bulb. This head contains a liquid that partially fills a glass bulb. Since the space is only partially filled, the glass bulb also has a trapped air bubble. As the temperature increases, the liquid expands and the air bubble is compressed. The pressure now rises rapidly and the bulb shatters, opening the cap. The size of the bubble (the amount of liquid) controls the operating temperature of the head.

The temperature at which the sprinkler head will operate varies with the installation. A head located over a boiler will need a higher operating temperature than one in a hotel hallway. The temperature rating of the head is indicated on the link and also by the color of the frame (table 5). When replacing sprinkler heads that have opened, be sure that a head set for the correct temperature is installed.

When inspecting sprinkler installations, be sure to check to see that the head has not been painted or covered in any way. The covering will raise the operating temperature or may even make the head inoperable.

TABLE 5. Color Indications for Sprinkler Operating Temperatures

Ceiling temperatures °F	Non-solder Sprinkler °F	Solder Sprinkler °F	Color
100	135-150	155-165	Bronze
150	175	212	White
225	250	286	Blue
300	325	360	Red
375	400		Green
475	500		Orange

Water supply

The sprinkler water supply systems are the same as the standpipe supply systems. However, the water supply is only calculated to be able to supply a few heads. Should a large volume of fire open many heads simultaneously, the supply system would be overwhelmed. For this reason, it is very important for a pumper to connect to the sprinkler system and supplement the supply. In addition, if the sprinkler valve is accidentally left closed, the use of the fire department siamese will bypass the closed valve and feed the system.

Fireground operating procedures

1. Pre-fire planning and inspection are important to locate and identify the sprinkler systems in the area. The nearest water supplies to the system should also be determined.

2. One of the first-arriving companies should lay one of the first lines from supply pumper to the sprinkler siamese. A second line should be laid if the officer feels that the flow will be high.

3. Remove the cap on the siamese intake. Some caps are removed with spanner wrenches, while others require that the cap be broken. Check the female swivel in the intake to make sure there is a gasket and that it turns.

4. Connect a line from the pumper to the siamese. Charge the system only when advised to do so by the officer.

5. Charge the sprinkler system.

6. When the fire is controlled, the opened sprinkler heads can be temporarily shut with a wooden wedge or a sprinkler shutoff device. If a large number of heads have fused, shut the control valve and open drain. All the time the control valve is shut, station someone at the valve with a means of communication so that the water can be immediately turned on in an emergency.

7. Replace the fused heads with new ones of the same temperature rating.

8. If the system is normally wet, shut the drain and open the control valve. If the system is a dry pipe, be sure that a watchman remains on the premises to open the control valve in case of another fire. The watchman should remain until the dry system has been reset by a competent mechanic. Note: Dry pipe valve must be taken apart to be reset.

Engine pressures

One of the most difficult jobs for a pump operator is to supply an operating sprinkler system. There is almost no way of determining how many heads have fused, so the flow is unknown. In addition, each particular installation will have many 90-degree bends and unknown lengths of pipe, so that estimating friction loss will be impossible.

For this reason, a rule of thumb that should supply most configurations has been developed. The figure used for supplying the system is 150-psi discharge at the pumper. If the pumper is unable to maintain the 150 psi, then too many heads have fused and an additional pumper will have to help supply the system.

Flow from a sprinkler head can be calculated if the pressure at the opening is known. The formula is $Q = 1/2P + 15$, where Q is the flow in gpm, P is pressure at the orifice in psi and 15 is a constant.

Example: What is the flow from a sprinkler head with a head pressure of 15 psi?

Step 1. Select the proper equation:

$$Q = \frac{1}{2}P + 15$$

Step 2. Determine the formula values:

$$P = 15 \text{ psi}$$

Step 3. Solve the equation:

$$\begin{aligned} Q &= \frac{1}{2}(15) + 15 \\ &= 7.5 + 15 \\ &= 22.5 \text{ gpm} \end{aligned}$$

Chapter 19

Tanker and Portable Pump Operations

Rural fire departments constantly face the problem of inadequate water supplies for firefighting. While a relay is one possible answer, the extra number of pumpers needed and the time it would take to reach the fire scene eliminates this as a solution. One way to move water from the source to the fireground is with tankers or portable pumps.

Tankers come in all shapes and sizes, from a pumper with a 1000-gallon tank to a converted 2000-gallon milk truck to a 6000-gallon converted gasoline tractor-drawn delivery truck (figures 1A and 1B). To use these tankers efficiently, the operator must known construction details, rating methods and operating procedures.

Figure 1A. Tanker/pumper combination.

Figure 1B. Tractor-drawn tanker.

Tanker construction

The ability of a tanker to load water at the source, carry it safely to the fireground, dump it, and then return for another load depends on the way it is built. The factors that influence tanker usability are:

Chassis — The chassis must be sufficient to carry the intended load.

Tank size — The tank must be able to supply a sufficient amount of water for the usual fires encountered in the district. The amount of water and the availability of tankers from neighboring fire departments are other factors to consider when determining the tank's capacity.

Baffles — Baffles are installed in the tanks to prevent the surging (sloshing) of the water as the tanker moves. The baffles must also have a number of openings between them, so that the entire load of water can be dumped quickly.

Vents — As water enters the tank, a large volume of air must be displaced. Vents eliminate the air and prevent an excessive build-up of pressure during filling, when the flow rate of air can reach 60 cubic feet per minute.

Intake — Tankers can be filled from either the top or the bottom. If a top fill is chosen, the line must be held in place. This is difficult if flows are higher than 300 gpm. In addition, turbulence and venting problems are created when using a top fill. If the fill is located near the bottom of the tank, turbulence is reduced and venting at the top of the tank can be accomplished. The only drawback of this type of fill is the back pressure created as the level of water in the tank increases. However, if a pump is used at the source, the back pressure is easily overcome.

Discharge — The discharge of the tanker should be large enough so that the entire contents of the tanker can be quickly dumped. This usually means that at least a 4½-inch pipe and gate are necessary (figure 2).

Figure 2. Tanker discharging water.

Pump — Pumps for tankers range from 1000 gpm to small portable pumps. Some tankers do not have pumps, but rely on gravity feed. The pump size and style chosen for the tanker depends on the needs of the individual fire department.

Water level — A water level indicator is a very important feature on the tanker. It will prevent overfilling which can cause icing conditions or soft ground. It will also indicate when the water supply is nearly exhausted, so that the next tanker can be moved in, or if no additional water is immediately available, provide the officer with time to make another strategy change.

Hose beds — The use of a tanker by a fire department indicates that water sources are not readily available. If water is not available, then only the hose

TANKER AND PORTABLE PUMP OPERATIONS

needed by the tanker at the fire scene for attack purposes is necessary. This means that perhaps two 200-foot, 1½-inch lines and one 250-foot, 2½-inch line are needed. Tanker shuttling requirements will add a few hundred feet of extra hose for filling and discharging. This means that the hose beds on a tanker can be small. Note that this applies only to tankers, since relay pumpers will need considerably more hose.

Tanker rating

A tanker cannot be rated simply in gallons per minute as a regular pumper, because the travel distance from source to fireground will vary, the speed of dumping will change, and the intake configuration is different for each tanker. To provide a standard for comparison, tankers can be compared by rating them in gallons per minute of water delivery per mile. This means that if it takes five minutes for a 1500-gallon tanker to go 1 mile to the source, fill and return to the fireground, dump the water, and return to the source, the tanker would be rated

$$\frac{1500 \text{ gallons}}{5 \text{ minutes} \times 2 \text{ miles (round trip)}} = 150 \text{ gpm per mile}$$

The first step in determining the rating of a tanker is to accurately compute the holding capacity of the tank. Table 1 outlines the steps necessary for calculating tank capacity.

TABLE 1. Tanker Capacity Calculation

Fire Department _____

Apparatus Designation _____

Rated Tank Capacity _____ gallons

A. Full loaded weight		(1) _____
B. Weight after dumping or discharging usable water		(2) _____
C. Empty weight taken after opening all drains and removing all possible water from piping and pump		(3) _____
D. Usable Capacity	$\frac{(1) - (2) \text{ pounds}}{8.35 \text{ lbs./gal.}} =$	_____ gallons
E. Maximum Capacity	$\frac{(1) - (3) \text{ pounds}}{8.35 \text{ lbs./gal.}} =$	_____ gallons

Once the capacity is determined, the measurement can begin. Place a pumper at a fill site. At the discharge site, prepare two 2½-inch hand lines for the tanker to connect to and a number of 55-gallon drums.

The time starts when the empty tanker stops at the fill site and continues as the tanker travels to the discharge site and dumps its water. When the tanker returns to the starting point, the time stops. The number of drums filled are counted and multiplied by 55 to determine the number of gallons actually delivered. Table 2 provides a convenient format for recording the necessary data.

Operation from a pumper tank

Using the water contained within the tank on the apparatus leads to a quick initial attack. The operation procedures are the same as for hydrant operations, except the tank to pump valve is opened to permit water to flow to the pump.

When operating from the tank, the operator must be continually aware of the length of time (at the current flow) that water will be available. Then, arrangements must be made for additional supplies either by relay or tanker shuttle. Table 3 lists the length of time various flows will last.

TABLE 2. Tanker Shuttle Capacity Calculation

Fire Department _____

Apparatus Designation _____

Total Run __1 mile_____

	Run 1 Time Sec.	Run 2 Time Sec.	Run 3 Time Sec.	Average Time Sec.
1. Tanker wheels stop at fill site. Timing starts				
2. Water flows from fill truck				
3. Water stops flowing				
4. Tanker wheels start moving from fill site				
5. Tanker wheels stop at dumping site				
6. Water starts flowing				
7. Water stops flowing				
8. Tanker leaves empty				
9. Tanker returns to fill site. Stop timing				
10. Total time (A)				
11. Number of drums filled				
12. Drums × 55 gallons (B)				
13. Tanker rating = $\frac{B}{A}$				

TABLE 3. Time Available for Water Flow from Tanks

Tank capacity (gallons)	Flow (gpm)			
	25	100	200	250
300	12 min.	3 min.	1½ min.	1⅕ min.
400	16	4	2	1⅗
500	20	5	2½	2
600	24	6	3	2⅖
700	28	7	3½	2⅘
800	32	8	4	3⅕
900	36	9	4½	3⅗
1000	40	10	5	4

When preparing apparatus specifications, it is important to remember that the piping from the tank to the pump must be capable of flowing 250 gpm in tanks up to 800 gallons and 500 gpm in tanks that hold 800 gallons or more.

One of the problems in operating from an apparatus tank is the transition from the tank to a hydrant. Because a higher rpm is necessary to deliver water from the tank, opening the hydrant intake without a throttle adjustment will add increased pressure to the discharge lines.

The following steps will provide for safe operation:

1. Open the tank to pump valve and adjust the throttle to obtain the correct discharge pressure.

2. Connect to the hydrant supply or the relay pumper (pressure source).

3. Open the intake supply valve half way, adjust the throttle control to the correct pressure, and close the tank to pump valve half way.

4. Open the intake supply valve completely, adjust the throttle control to the correct pressure, and close the tank to pump valve fully.

5. Open the tank fill valve to refill the tank while water is being discharged. The operator must ensure that the top cap or vent on the tank is opened, otherwise the large amount of water flowing into the tank will not have an outlet. This will cause increased pressure that can rupture or bulge the tank.

Operation directly from a tanker

When the fire is located on a narrow road or on an unpaved road, shuttling tankers in, having them dump water, and then turning them around to refill proves difficult. In such a situation, a pumper can lay a supply line from the hard surface roadway to the fireground with a siamese having clapper valves on the end (figure 3).

A tanker connects to the siamese and pumps to the attack pumpers. The clapper valve prevents the backward flow of water out the other intake. When the first tanker runs out of water, the line is disconnected, the clapper valve closes, and the second tanker begins pumping. While tanker 2 is pumping, tanker 1 can refill. The length of time required for refilling will determine how many additional tankers are needed.

By keeping the tankers away from the fire scene, road time is substantially reduced. However, this method will work only if the tanker has a pump sufficient to supply the required volume of water to the attack pumper.

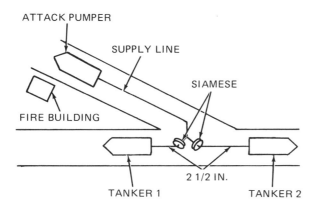

Figure 3. Operation directly from a tanker.

Operations from portable tanks

A big assist to tanker operations is a folding tank, which serves as a reservoir for a tanker shuttle. This allows a tanker to dump its entire load of water immediately and return to the source of water for refilling. This also points out the need for a large dump valve to permit quick emptying. An attack pumper at the fireground drafts water from the folding tank. Two types of tanks are shown in figures 4A and 4B.

To set up operations from a portable tank, the following procedures can be used:
1. The attack pumper goes to the fire scene and begins operation from the internal tank.
2. The folding or collapsible tank is set up in a position that will permit easy access for the tanker dump valves and from which the attack pumper can easily draft.
3. A hard sleeve with strainer is attached to the attack pumper and placed in the basin.
4. A tanker with a short length of hose attached to the dump valve backs up to the basin.
5. The tank vents are opened and the dump valve is opened.
6. When the tanker is empty, it returns to the fill point where a pumper is ready to begin the refilling operation.
7. As the water in the basin is used, another tanker backs up and discharges its water.
8. The number of tankers needed depends on the carrying capacity, distance, refill time, and fire flow requirements.

Folding tanks can also be used as the water reservoir in remote areas (figure 5A and 5B).

A. Collapsible type

B. Folding type

Figure 4. Folding Tanks

Portable pump operation

The use of portable pumps for the fire service has never been fully explored. They can be used for fireground operations when the apparatus either cannot get near the water source or the fire is inaccessible to a regular pumper. In addition, these pumps can be used for salvage operations by removing water from basements.

What exactly is a portable pump? NFPA 191 classifies portable pumps as:

1. Small volume, high pressure	15-20 gpm at 200 psi	175 pounds overall weight
2. Medium volume, medium pressure	50 gpm at 90 psi 100 gpm at 60 psi	150 pounds 150 pounds
3. Large volume, low pressure	100 gpm at 50 psi 250 gpm at 20 psi	150 pounds 150 pounds
4. Medium volume, high pressure	90 gpm at 250 psi 160 gpm at 90 psi	200 pounds 200 pounds
5. Extra-large volume, medium pressure	500 gpm at 100 psi	200 pounds

While these five general classifications allow wide variations in portable pump design, the definition can be generalized as follows: Can be carried by two men, volume delivery for supplying pumper, pressure delivery to operate attack lines, protective frame (figure 6).

Figure 7 shows a floating portable pump, with a built-in strainer on the pump intake, being used to supply a 1½-inch line directly. Figure 8 illustrates a portable pump drafting water from an inaccessible source and directly supplying an attack line.

A. Airlifting tank to fire scene

B. Filling tank from helicopter drop.

Figure 5. Folding tanks in remote areas.

Figure 6. Portable pump.

Figure 7. Floating portable pump.

Figure 8. Portable pump at inaccessible static source.

Portable pumps in relay

A portable pump is used to relay water when a pumper cannot be positioned at the water source. The procedures for establishing the relay are:

1. The attack pumper stretches a supply line from as close as possible to the water source to the fireground (figure 9). The portable pump is left at the water source.

2. Operation begins by using the water from the tank.

3. The portable pump is carried to the static source and placed in operation. The supply line lay is completed and the hose is connected to the discharge of the portable pump.

4. Water is supplied to the attack pumper under pressure. The changeover from tank to portable pump should be accomplished as outlined in the section "Operations from portable tanks."

5. A firefighter should stay at the water source to ensure that the water delivery continues uninterrupted.

Portable pumper for supplying attack lines

A portable pump is used to directly supply the attack lines when the pumper cannot reach the fire building and when the water source is close to the involved building. The procedures for establishing the operation are:

1. The portable pump is carried to the static source and used to draft water (figure 10).
2. Hand lines are connected to the discharge and advanced to the fireground.
3. If the distance is too great, additional portable pumps may be used in relay.
4. A firefighter should remain at the water source to ensure that a continued supply will be available.

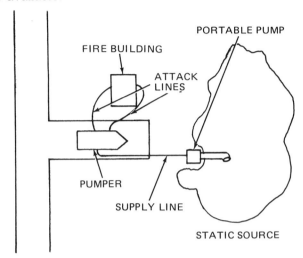

Figure 9. Portable pump operation in a relay.

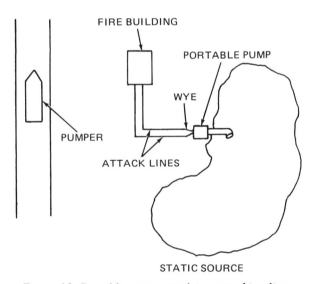

Figure 10. Portable pump supplying attacking lines.

Chapter 20

Testing and Maintenance

In 1911, the International Association of Fire Engineers (now the International Association of Fire Chiefs) conducted some brief tests on the newly developed gasoline-driven pumper. The following year the National Board of Fire Underwriters (now the American Insurance Association) joined with the fire engineers to conduct the tests at their annual meeting. The National Fire Protection Association then joined the group testing pumpers. The NFPA Committee on Automotive Apparatus continues to set the testing standard for pumpers.

In May 1965, the AIA discontinued its program of testing fire department pumpers, and Underwriters' Laboratories, an independent, not-for-profit, engineering laboratory, took over the pumper testing program, which they continue to perform.* Any testing agency chosen by the purchaser may do this testing, provided it is acceptable to the state. Two major types of tests are performed, the certification test and the service test.

An acceptance test is conducted at the factory. Section 11-2.1 of NFPA 1901 describes this test.

"11-2.1 Tests Performed by the Manufacturer.

"11-2.1.1 The fire pump on the completed vehicle shall be thoroughly run in by the manufacturer before being delivered, by operating for a minimum of two hours at draft delivering rated capacity of 150 psi net pump pressure for at least 1 hr., 70 percent of rated capacity at 200 psi net pump pressure for at least ½ hour and 50 percent of rated capacity at 250 psi net pump pressure for at least ½ hour.

"11-2.1.2 All pumps shall by hydrostatically tested by the pump manufacturer for 10 min. at a pressure not less than 350 psig.

"11-2.1.2.1 Two-stage series-parallel pumps shall be tested either hydrostatically or hydrodynamically by the pump manufacturer at a discharge pressure of 500 psig."**

An affidavit to the effect that this test has been conducted is given to the Underwriters Laboratories or the testing agency chosen by the fire department. They then witness a certification test as outlined in NFPA 1901-11.2.2. If the apparatus satisfactorily completes the test, the testing agency will issue a cer-

---

*"Guide for Certification of Fire department Pumpers," Underwriters' Laboratories, Inc., 333 Pfingsten Road, Northbrook, Ill. 60062.

**Section 11.2.1 Tests Performed by the Manufacturer. Reprinted with permission from NFPA 1901-1979, Standard for Automotive Fire Apparatus. Copyright © 1979, National Fire Protection Association, Quincy, Massachusetts 02269. This reprinted material is not the complete and official position of the NFPA on the referenced subject, which is represented only by the standard in its entirety.

tification stating that full compliance with the pumping standards has been met (figure 1).

Manufacturers do not automatically run an acceptance test, so that it must be a requirement of the specifications which the department puts out for bid.

Once the apparatus has been delivered, the department should run its own test in accordance with the National Fire Protection Association Standard 1901.*** This is the same standard that UL uses as a basis for the certification test.

The certification test is summarized in the following paragraphs. The text of the standard follows the summary.

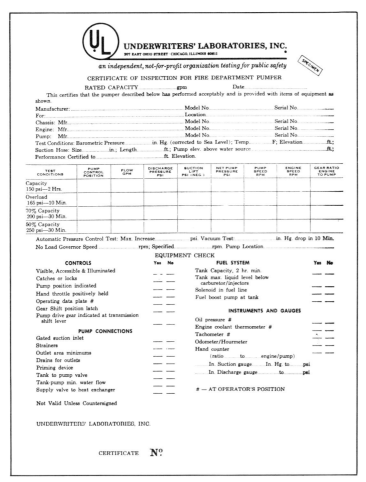

Figure 1. Underwriters' Laboratories test certificate.

Certification test and summary

Flow

100% of rated capacity	150 psi net pump pressure	2 hours
70% of rated capacity	200 psi net pump pressure	½ hour
50% of rated capacity	250 psi net pump pressure	½ hour
100% of rated capacity	165 psi net pump pressure	10 minutes

***Standard for Automotive Fire Apparatus, NFPA 1901, National Fire Protection Association, Batterymarch Park, Quincy, Mass. 02269.

Automatic pump Pressure Control

Draft	150 psig	100% capacity	With the discharges closed slowly, the discharge pressure cannot exceed 180 psig.
Draft	150-90 psig	100% capacity	The throttle is shut to reduce discharge pressure to 90 psig. With the discharges closed slowly, the discharge pressure cannot exceed 120 psig.
Draft	250 psig	50% capacity	With the discharges closed slowly, the discharge pressure cannot exceed 280 psig.

Pump Priming Device

0-1000 ft altitude	capped 20-ft length of hard sleeves	discharges uncapped	22 in of vacuum	At the end of 10 min the reading must be at least 12 in.
1001 - 2000 ft altitude	capped 20-ft length of hard sleeves	discharges uncapped	21 in of vacuum	At the end of 10 min the reading must be at least 11 in.
2001 - 3000 ft altitude	capped 20-ft length of hard sleeves	discharges uncapped	20 in of vacuum	At the end of 10 min the reading must be at least 10 in.

For each 1000-foot increase in altitude, the required vacuum is reduced 1 inch. During the 10-minute period of the test, the primer cannot be operated.

NFPA standard 1901 certification test

The certification test performed by the testing agencies is conducted in accordance with NFPA 1901. The text reproduced below is taken from the 1979 (the current) edition of the standard.

"11-2.2 Certification Tests.

"11-2.2.1 A three-hour certification test shall be performed and shall consist of drafting water and pumping rated capacity against a net pump pressure of 150 psi for a continuous period of two hours, followed by two ½-hour periods of continuous pumping, during one of which at least 70 percent of the rated capacity shall be delivered at net pump pressure of 200 psi and during the remaining ½ hour, 50 percent of the rated capacity shall be delivered at a net pump pressure of 250 psi.

"11-2.2.2 The apparatus shall be subjected to a 10-minute overload test to demonstrate its ability to develop 10 percent excess power (see 3-1.3.2). The test shall consist of discharging rated capacity from draft at 165 psi net pump pressure.

"11-2.2.3 A test shall be conducted to check the performance of the automatic pump pressure control (see 3-3.2.7) pumping from draft and discharging rated capacity at 150 psig pressure. With the pressure control set in accordance with manufacturer's instructions, after all outlet valves have been closed slowly, pump discharge pressure shall not have increased more than 30 psi from its original value.

"11-2.2.3.1 After the 150 psig test specified in 11-2.2.3 has been completed, the discharge shall be reestablished at rated capacity at 150 psig pressure. Pressure shall then be reduced to 90 psig by using the throttle. With the pressure control set in accordance with manufacturer's instructions, after all outlet valves have been closed slowly, pump discharge pressure shall not have increased more than 30 psi from its original value.

"11-2.2.3.2 After the 90 psig test specified in 11-2.2.3.1 has been completed, the discharge shall be established at 50 percent of rated capacity at 250 psig pressure. With the pressure control set in accordance with manufacturer's instructions, after all outlet valves have been closed slowly, pump discharge pressure shall not have increased more than 30 psi from its original value.

"11-2.2.4 A vacuum test shall be performed and shall consist of subjecting the interior of the pump, with capped suction and uncapped discharge outlets, to a vacuum of 22 in. Hg by means of the pump priming device. The vacuum shall not drop more than 10 in. Hg in 10 minutes. The primer shall not be used after the 10-minute test period has begun. (See 3-3.2.9.)

Exception: At altitudes above 1000 feet, the value of required vacuum shall be reduced by 1 inch per 1000 feet.

"11-2.2.5 Tests shall be made with all accessories and power-consuming appliances attached.

"11-2.2.6 The tests shall be stopped only as necessary for changing hose and nozzles. During and after the tests, the engine, pump transmission, and all parts of the machine shall show no undue heating or excessive strain or vibration; and the engine shall show no loss of power, overspeed, or other defect.

"11-2.2.7 For pumpers rated less than 1500 gpm capacity, tests shall be made with 20 feet of hard suction hose in a single line. For pumpers rated 1500 gpm or more, two 20-foot suction lines may be used. External and internal strainers shall be used on all tests. (See 3-2.2.1.)

"11-2.2.8 Where a water tank of 300 gallons or larger is furnished, a test shall be performed to check the tank to pump flow rate as specified in 6-1.2.3 of this Standard. The required rate of flow shall be maintained throughout the discharge period until at least 80 percent of the rated capacity has been discharged. (See 6-1.2.3.)"†

Net pump pressure calculations

The amount of work that a pump must perform to deliver rated flow depends on whether the water enters the eye of the impeller under pressure or whether the pump has to work to bring the water up into the eye (drafting). When water enters the impeller under pressure, the pump has less trouble building up additional pressure. Under draft conditions, the pump must work to create and maintain the vacuum that brings the water into the strainer, up the drafting sleeve, and into the impeller.

Net pump pressure is defined as:

$$\text{NPP} = \frac{\text{psig} + \text{lift (ft)}}{2.3} + \frac{\text{suction loss (ft)}}{2.3}, \text{ where}$$

NPP = net pump pressure

psig = discharge pressure at the gage

lift = vertical lift in feet

suction loss = loss in strainer and hard sleeves in ft (table 1)

2.3 = conversion factor from ft to psi

†Section 11-2.2 Certification Tests. Reprinted with permission from NFPA 1901-1979, Standard for Automotive Fire Apparatus, Copyright© 1979, National Fire Protection Association, Quincy, Massachusetts 02269. This reprinted material is not the complete and official position of the NFPA on the referenced subject, which is represented only by the standard in its entirety.

PUMP OPERATORS HANDBOOK

Since the gage pressure at the discharge is the quantity needed for the test, with the net pump pressure being known, the equation can be transformed to

$$\text{psig} = \text{NPP} - \frac{\text{lift}}{2.3} - \frac{\text{suction loss}}{2.3}$$

Now, it can be easily seen that the net pump pressure is less than the gage pressure for the drafting situation. By using this formula, the discharge pressure is independent of the lift. This equalizes the test no matter where it is performed.

Example: What gage pressure is required for a 1000-gpm pumper at the 100 percent capacity test, operating from an 8-foot lift, with 20 feet of 4½-inch hard sleeve?

Step 1. Determine the correct formula:

$$\text{psig} = \text{NPP} - \frac{\text{lift}}{2.3} - \frac{\text{suction loss}}{2.3}$$

Step 2. Determine the formula values:

$$\text{NPP} = 150 \text{ psi } (100\% \text{ capacity test})$$

$$\frac{\text{lift}}{2.3} = \frac{8}{2.3} = 3.5 \text{ psi}$$

$$\frac{\text{Suction loss}}{2.3} = \text{table 1}$$

First 10 feet = 12
Second 10 feet = 2½

$$= \frac{12 + 2½}{2.3}$$

$$= 6.3 \text{ psi}$$

Step 3. Solve the equation:

$$\text{psig} = 150 - 3.5 - 6.3$$
$$= 140.2 \text{ psi}$$

Note that the suction allowance must be reduced by 1 psi for the 70 percent test and 2 psi for the 50 percent capacity test.

Example: What gage pressure is required for a 1250-gpm pumper at the 50 percent capacity test, operating from a 9-foot lift, with 20 feet of 5-inch hard sleeve?

Step 1. Determine the correct formula:

$$\text{psig} = \text{NPP} - \frac{\text{lift}}{2.3} - \frac{\text{suction loss}}{2.3}$$

TESTING AND MAINTENANCE

Step 2. Determine the formula values:

$$NPP = 250 \text{ psi } (50\% \text{ capacity test})$$

$$\frac{\text{lift}}{2.3} = \frac{9}{2.3} = 4.0 \text{ psi}$$

$$\frac{\text{suction loss}}{2.3} = \text{table 1}$$

First 10 ft = 12 ½
Second 10 ft = 2
Reduced allowance for 50% test = 2

$$\text{Loss} = \frac{12½ + 2}{2.3} - 2 = 4.3 \text{ psi}$$

Step 3. Solve the formula:

$$\begin{aligned} \text{psig} &= 250 - 4.0 - 4.3 \\ &= 241.7 \text{ psi} \end{aligned}$$

Pump service test

"The fire apparatus driver/operator shall identify the percentages of rated capacity, rated pressures, and the capacity in gallons per minute at the rated pressures a fire department is designed to deliver.

"The fire service apparatus driver/operator, given a fire department pumper and the necessary equipment, shall demonstrate an annual pumper service test.

"The fire apparatus driver/operator shall identify the following conditions that may result in pump damage or unsafe operation, and identify corrective measures:

"(a) Cavitation (see Chapters 15 and 16)
"(b) Leaking fuel, oil, or water
"(c) Overheating
"(d) Unusual noises
"(e) Vibrations
"(f) Water hammer (see Chapter 12)." ††

While performance of a new pumper is certified by Underwriters' Laboratories or other testing agencies, when it is delivered, the pumper must continue to operate effectively for 15 or more years. It must be reliable on the fireground. Time lost while an additional pumper is brought in to replace a broken-down piece of apparatus can cause tremendous losses in life and property.

A schedule for testing pumps thus becomes an extremely important item for all fire departments. These tests should be performed at least once a year. They will help to locate defects that could otherwise go unnoticed or unsuspected since the pump is rarely called upon to produce at full capacity.

Two types of service tests are conducted on a regular basis. The pump service test determines the ability of the pump to deliver its rated quantity at rated pressures. The dry prime test checks the ability of the pump to develop a vacuum.

---

††Paragraphs 3-1.4, 3-1.5, and 3-1.6. Reprinted with permission from NFPA 1002-1982, Standard for Fire Apparatus Driver/Operator Professional Qualifications, Copyright© 1982, National Fire Protection Association, Quincy, Massachusetts 02269. This reprinted material is not the complete and official position of the NFPA on the referenced subject, which is represented only by the standard in its entirety.

The pump service test consists of drafting with a lift of not more than 10 feet and pumping at a net pump pressure as follows:

100% rated capacity	150 psi net pump pressure	20 minutes
70% rated capacity	200 psi net pump pressure	10 minutes
50% rated capacity	250 psi net pump pressure	10 minutes

There are three variable factors that are interrelated with this test: pump speed, pump pressure, and pump discharge. This interrelationship is seen by increasing the pump rpm which will increase the pump pressure or the pump discharge or both. Pump speed is controlled by the engine throttle, pump pressure by the hose layout and position of the discharge gate, and flow by nozzle size. The only way to reach the test conditions is by adjusting all three variables.

Test layout

The discharge hose layouts consist of 2½-inch supply lines connected to a smooth bore nozzle. The hose acts not only as a water carrier, but it also supplies some friction loss that is required to reduce engine discharge pressure to the required nozzle pressure. Since short lengths of hose are convenient to handle, the additional friction loss necessary is provided by partially closing a discharge gage.

For example, to deliver 1000 gpm through a 2-inch tip at 250-psi net pump pressure requires a nozzle pressure of 71 psi. This means that 79 psi of friction loss must be introduced. If 100 feet each of two 2½-inch lines are used, the friction loss in each line flowing 500 gpm will be 55 psi. The remaining 24-psi friction loss necessary for the desired nozzle pressure will come from partially closing the discharge gate.

The test setup for various size pumps is shown in table 2.

Test equipment

To accurately test a pumper for performance, both a pitot gage and a pump pressure gage that have been carefully tested for accuracy are required. Gage testing can best be done with a dead weight gage tester. Smooth bore nozzles of accurate size should be used with the pitot gage. Nozzles should be used on a siamese deluge gun for greatest accuracy. A stream straightener behind the nozzle should also be used.

If a deluge gun is not available, a 2½-inch nozzle can be tied down for the test (figure 2).

A comparison of the revolutions per minute of the pump to deliver the same

Figure 2. Service test.

TABLE 1. Allowances for Friction Loss in the Suction Hose

Rated capacity of pumper (gpm)	Diameter of suction hose (inches)	Allowance (feet) For 10 ft. of suction hose	For each additional 10 ft. of suction hose
500	4	6	plus 1
	4½	3½	plus ½
750	4½	7	plus 1½
	5	4½	plus 1
1000	4½	12	plus 2½
	5	8	plus 1½
	6	4	plus ½
1250	5	12½	plus 2
	6	6½	plus ½
1500	6	9	plus 1
1500	4½ (dual)	7	plus 1½
1500	5 (dual)	4½	plus 1
1500	6 (dual)	2	plus ½
1750	6 (dual)	12½	plus 1½
1750	4½ (dual)	9½	plus 2
1750	5 (dual)	6½	plus 1
1750	6 (dual)	3	plus ½
2000	4½ (dual)	12	plus 2½
2000	5 (dual)	8	plus 1½
2000	6 (dual)	4	plus ½

Note: The allowance computed above for the capacity test should be reduced by 1 pound for the allowance on the 200 pound test and by 2 pounds for the allowance on the 250-pound test.

TABLE 2. Pump Service Test Setup

Capacity	2½" lines	Length ft.	Tip	NP	GPM	Net pump pressure
750-gpm pump						
100%	2	50	1-3/4"	68	750	150
70%	1	50	1-1/2"	62	525	200
50%	1	100	1-1/4"	66	375	250
1000-gpm pump						
100%	2	100	2"	71	1000	150
70%	2	100	1-3/4"	59	700	200
50%	1	100	1-1/2"	56	500	250
1250-gpm pump						
100%	4	100	two 1-1/2"	88	625 / 625	150
70%	2	100	1-7/8"	70	875	200
50%	2	100	1-1/2"	88	625	250
1500-gpm pump						
100%	4	100	two 1-3/4"	68	750 / 750	150
70%	2	100	2"	78	1050	200
50%	2	100	1-3/4"	68	750	250

Note: Insert a gate valve in the 2½-inch line for additional friction loss.

quantity of water will indicate the condition of the pump and any deterioration. Use a revolution counter and a stop watch to check the tachometer on the apparatus. The counter connection on the pump panel usually will indicate 1/10 or 1/2. This means that the tachometer reading obtained with the counter must be multiplied by 10 or 2 to get pump rpm.

Test procedure

1. Record the preliminary information on the pump test form (figure 3).
2. Position the apparatus for drafting.
3. Drain all water from the pump.

```
Apparatus _____ Pump Make _____ Date _____
Type of Pump _____ Model _____ Capacity _____
Gear Ratio _____ Engine Make _____ No. of Cylinders ___
Tested by _____
=================================================
                    PRIMING PERFORMANCE
Dry Prime — Initial Vacuum _____ inches
            After 10 minutes _____ inches
Draft — Lift _____ feet; _____ feet of hard sleeve
        Time to prime _____ seconds
=================================================
                CALCULATION OF GAUGE PRESSURE
PSIG = NPP − lift/2.3 − friction loss/2.3    NPP = _____

                                             lift/2.3 = _____

                                             fl/2.3 = _____

                                             PSIG = _____
=================================================
                        WATER FLOW
Gpm  Tip  Nozzle    Net Pump   Gauge     Time    Pump
          pressure  pressure   pressure  start/  rpm
                                         finish
     150
     200
     250
=================================================
                    PRESSURE CONTROL
Initial Pressure _____ Flow _____ Shut down pressure _____
=================================================
                    ENGINE CONTROLS
Oil Pressure _____ Water Temperature _____ Vibrations _____
Leaks _____
Remarks _____
```

Figure 3. Pump test record form.

4. Run the dry prime test as outlined in this chapter. Record the results on the form.

5. Connect two hard sleeves and the strainer together and then attach them to the pump intake. Attach a rope to the strainer.

6. Hammer all couplings together to ensure an airtight seal.

7. Lower the strainer end of the hard sleeve into the drafting source.

8. Connect the hose and appliances as indicated in table 2.

9. Start the priming mechanism and time the length of time necessary to obtain the prime. A prime should be obtained in less than 45 seconds. If a prime is not obtained, check for a malfunction. Record the results on the chart.

10. Run the pump for a few minutes discharging some water, so that both the pump and engine can warm up.

11. Calculate the necessary gage pressure, taking into account the size of the hard sleeves, the height of the lift, and the length of the hard sleeve.

12. Keep one or both of the discharge gates partially closed before building up the pump pressure. Increase engine rpm until the required gage pressure is obtained.

13. Check the nozzle pressure with a pitot gage.

14. If the pump pressure is correct but the nozzle pressure is too high, close the gate a little to introduce more friction loss. This will lower the nozzle pressure but it will increase pump pressure. Correct the pump pressure by shutting the throttle a little. Check the pitot reading again at the nozzle.

15. If the pump pressure is correct, but the nozzle pressure is too low, a discharge gate must be opened slightly. This will reduce pump pressure and in-

TESTING AND MAINTENANCE

crease nozzle pressure. The throttle must again be adjusted for correct discharge pressure.

16. When the pump pressure and discharge quantity are satisfactory, continue the test for the required length of time. If the time taken to make the initial adjustments is not too great, the test time can be counted from the first build-up of pressure. Check the gates to be sure that they do not vibrate open or closed. Record the results on the chart.

17. Readings should be taken with the pitot gage every five minutes with the 20-minute test and every two minutes with the 10-minute test.

18. After running the 100 percent capacity test, the two pressure tests follow immediately. Do not drop the water between these tests. The only time interval is that which is necessary to change nozzle size and layout.

19. The transfer valve is set to pressure for the 200-psi test.

20. Adjust the discharge as described in steps 11 through 17.

21. After completing the 200-psi test, change the nozzle and hose layout for the 250-psi test. Keep the transfer valve in the pressure position.

While conducting the annual service test, discharge gates are partially closed to either create friction loss or to achieve a higher pump pressure. This, however, decreases the flow. On the fireground, it is not beneficial to increase pump discharge pressure at the expense of reduced flow through partially closing a discharge gate.

The only time a discharge gate should be partially closed on the fireground is during the use of multiple lines of different diameters that require different flows. Then, the pump pressure is set for the highest pressure required and the other lines are gated back to reduce flow. The discharge gates are closed until the proper pressure is indicated on the individual discharge gage.

Dry prime test

The dry prime test provides information concerning the tightness of the pump and the priming system. It indicates if the pump is in good condition for drafting. This test is also used to check the inner liner of the hard sleeve.

The hard sleeve has a rubber lining, fabric, a rubber cover, and a helix of wire set in rubber between reinforcing layers of fabric. If the hard sleeve is strained or bent, it is possible for the layers of the lining to separate. Then, when drafting under vacuum conditions, the liner will collapse and prevent water from entering the pump. When the pressure returns to atmospheric, the liner returns to normal. For this reason, the sleeves are checked during the priming test.

The test procedure is as follows:

1. Insert a lit flashlight facing away from the pump panel in one end of the hard sleeve.
2. Connect one section of hard sleeve to the intake of the pump. Make sure the flashlight is facing out. Support the hard sleeve in a level position (figure 4).
3. Cap the hard sleeve with a piece of Plexiglas with a gasket of the correct size for the sleeve being used.
4. Drain the pump and make sure the tank-to-pump valve is closed.
5. Close all pump openings.
6. Operate the primer until a reading of 22 inches is obtained.
7. Stop priming and shut off engine.
8. Begin timing.
9. Look into the hard sleeve to make sure that the inner lining does not collapse under vacuum.
10. Listen for air leaks which are sometimes audible with the engine stopped.

PUMP OPERATORS HANDBOOK

11. The vacuum reading should not fall below 12 inches in less than 10 minutes.

12. If the vacuum is lost, it means that there is an air leak in the system. Check the drains, gates, connections, packing, and relief valve for leaks.

13. Connect the other hard sleeve and repeat steps 1 through 9.

14. At the conclusion of the test, check the priming reservoir for oil if a rotary gear primer is used.

Figure 4. Dry prime test.

TRUCK MAINTENANCE

To operate efficiently, the fire department apparatus must be well maintained. If a truck fails to start, breaks down on the way to a call, or fails to operate on the scene, there can be disastrous results. The driver/operator is responsible for ensuring that the following areas are in a correct and safe operating condition. A suggested check list is shown in table 3.

"The fire apparatus driver/operator shall demonstrate the performance of routine tests, inspections, and servicing functions required to assure the operational status of fire department pumpers, including:

"(a) Battery check
"(b) Booster tank level (if applicable)
"(c) Braking system
"(d) Coolant system
"(e) Electrical system
"(f) Fueling
"(g) Lubrication
"(h) Oil levels
"(i) Pump
"(j) Tire care
"(k) Tools†††

Remember, the driver/operator is not the mechanic. So, although these maintenance suggestions can be easily performed, the more complex maintenance and repairs should be referred to the fire department mechanic. Any leaks, vibrations, or unusual operating problems need to be checked immediately.

Battery and electrical system

The electrical system of a fuel pumper consists of the battery or batteries, the lighting system, the alternator, the voltage regulator, and any electrical motors.

*Paragraphs 3-1.1, 3-1.4, 3-1.5, and 3-1.6. Reprinted with permission from NFPA 1002-1982, Standard for Fire Apparatus Driver/Operator Professional Qualifications, Copyright©1982, National Fire Protection Association, Quincy, Massachusetts 02269. This reprinted material is not the complete and official position of the NFPA on the referenced subject, which is represented only by the standard in its entirety.

The electrical system should be checked as follows:
Battery - water level is correct
- terminals are clear of corrosion
- specific gravity is correct
Voltage regulator - if battery is at full charge and the ammeter shows a high charge, the voltage regulator needs to be checked.
Alternator - if the battery is uncharged and the ammeter shows little charge, the alternator and voltage regulator need to be checked.
Electrical equipment - the operation of all the lights, motors, and switches need to be checked to ensure correct functioning. Terminals need to be checked for corrosion and damage from moisture.

TABLE 3. Suggested Truck and Pump Maintenance List

Electrical System	**Braking System**	**Pump**
Battery	Master cylinder	Pump transmission
Voltage regulator	Air pressure	Pressure gages
Alternator	Air tank	Motor gages at panel
Emergency warning lights		Gates operate freely
Safety lights	**Tires**	Valves operate freely
Motors — booster reel	Visual check	Pump bearings
Siren/horn	Air pressure	Priming pump
Liquid Levels	Mountings	Transfer valve
Oil and oil filter		Gaskets and washers
Gasoline	**Equipment**	Check valves
Automatic transmission	Breathing apparatus	Suction threads
Power steering	Ladders	Packing
Coolant System	Cutting tools	Relief valve
Portable power tools	Nozzles	
Windshield washer	Hose	
	Hose appliances	

Liquid levels

There are many areas of the pumper containing liquid that need to be checked. The gages should be compared with a visual check of the liquid level to ensure that the gage is reading correctly. Additions should be made carefully and with the correct fluid. Those levels that need to be checked are:
Booster tank water level.
Coolant system.
Fuel tank.
Oil level.
Automatic transmission fluid.
Power steering fluid.
Windshield washer fluid.

Braking system

One of the most important safety items to be checked is the apparatus braking system. If the truck has air brakes, they can be tested by starting the truck and seeing how long it takes to get sufficient air pressure in the system. Any extensive delay means there is an air leak which needs immediate repair. In addition, failure to maintain air pressure needs immediate attention. Water should be bled from the air tanks. For mechanical brakes, the level of the fluid reservoir of the master cylinder should be checked.

One of the best methods for determining brake problems is the driver's sensivity to the changing feel of the truck. Gradual degradation is especially noticeable to a regular driver. This deterioration should be brought to the attention of the department.

Lubrication

One of the major items of protective maintenance is the lubrication of the pumper. In addition to maintaining the motor oil level, lubrication of all the metal-to-metal contacts is necessary. However, the proper grade and consistency of lubricant must be used to ensure that the parts remain protected as well as to reduce wear.

Tires

Each of the tires should be checked for the correct air pressure using a truck tire gage. In addition, each tire should be checked visually for wear, breaks, foreign matter in the treads, as well as for cracking due to age. Finally, the lugs must be checked to ensure they are tight.

Equipment maintenance

All removable equipment needs to be checked to ensure that it is ready for use. Items such as breathing apparatus must have the correct air pressure and be ready to use. Opening and cutting tools must be free of rust and have their working surfaces at the prescribed sharpness level. Nozzles must be free of dirt and have free-moving swivels and good gaskets. Hose must be clean with clean couplings and good gaskets. Portable tools with their own engines should have the correct fuel level, proper lubrication, and be ready to operate.

Pump maintenance

A good pump maintenance program will minimize downtime and provide for more reliable operation at an incident. Correct operation of the pump will also help to reduce the number of breakdowns. A suggested check list is shown in table 3.

One of the major problems encountered in pump breakdown is a result of the pump overheating. This results from the pump turning but no water flows through it. Indications of an overheated pump are: Pump overheat light goes on, steamer cap becomes hot, seizing of the clearance ring.

To avoid this condition, sufficient water needs to flow through the pump. To do this, open the tank-to-pump valve and the tank fill valve to allow the tank water to circulate. The following preventive maintenance functions should be performed by the driver/operator to ensure that the pump is ready to operate:

1. Main pump and drive unit — The pump bearings, drive unit bearings and all gears are supplied with oil from the drive unit housing. Use a good grade SAE 90 motor oil. Keep oil level between the high and low mark on the bayonet gage.

2. Priming pump — Keep the primer oil tank filled with SAE 30 motor oil.

3. Transfer valve — Operate the valve under pressure to ensure that the valve does not freeze in position. In addition check that the valve is greased and that any O rings are flexible.

4. Gaskets and washers — Inspect the gaskets and washers on the hard sleeves and the intakes to ensure they are in good condition. Poor gaskets can prevent the pump from getting a prime for drafting.

5. Check valves — The check valves should be free to swing. Ensure that no foreign matter is caught between the valve and the seat.

6. Suction threads — Cast iron suction threads should be coated lightly with grease once a month.

7. Packing — Packing is designed to be moist to remain soft and pliable. The water gets to the packing from the discharge side of the pump. Water should drip from the packing even when the pump is not engaged. The packing can be adjusted or replaced in accordance with the manufacturers' recommendations.

8. Relief valve — Operate the relief valve regularly to ensure that the parts move freely. In addition, ensure that any O rings remain flexible.

Chapter 21

Driver Training

Obviously, if a pumper and its personnel do not arrive on the scene, the safety of the public as well as the other firefighters is in jeopardy. An individual can be the most knowledgeable pumper operator in the department, but the operator must get the apparatus to the incident safely if it is to be useful.

In addition to knowing how to handle the pump, an individual who is selected to become an apparatus driver should have:

Good eyesight, be free from medical problems which would cause loss of vehicle control, and be in excellent physical condition.

Knowledge of the routine maintenance of the truck.

The ability to remain calm while responding under emergency conditions.

Knowledge of streets, response routes from other stations, and potential traffic problem areas.

Knowledge of potential water sources, particularly in areas that require the use of static water sources.

Safe driving

Unfortunately, in many fire departments the selection of the pumper driver/operator is based upon seniority or popularity. When selections are made this way, the tremendous responsibility of getting the vehicle to the scene in one piece may be overlooked.

When an alarm is received, the tendency is to try and get the firefighters and equipment to the scene as quickly as possible, disregarding traffic, weather, and road conditions. The truck weaves in and out, crosses over the center line to pass other vehicles, and ignores traffic control signals. This activity results in accidents, injuries, and jeopardizes the public safety. Yet, in many cases, the drivers who operate in this manner continue unchallenged by their superiors. Further, some departments even award these individuals special recognition in the mistaken belief that this is the correct way to drive emergency apparatus.

Drivers must remember that they are handling large and heavy vehicles. Many times the drivers are only familiar with handling the family car, which has grown smaller over the years. To suddenly switch to driving a big truck can create many problems.

Individuals driving a pumper on an emergency call must guard against being caught up in the excitement of the siren, lights and horns and resist throwing out caution. Most states only permit disregarding traffic laws when the public will not be endangered. This means the driver must make certain that an intersection is clear before entering, even if the traffic light gives the apparatus the right of way.

Drivers must remember that some individuals will stop immediately upon hearing a siren; they may freeze with fright and not pull over. Other traffic may then have to stop and the apparatus will not have any place to go and an accident could result. In addition, with total climate control for passenger cars, windows stay closed in summer and winter, and many drivers play radios and tapes, so they may not notice emergency apparatus until it is too late. The apparatus driver must continually be aware of the traffic in front, indications of movement to the left or right, the sudden decision of a car in front to make a left turn, and oncoming traffic.

The driver/operator must also keep in mind the response routes of other apparatus responding on the call. In many cases, the warning devices make it impossible to hear other emergency apparatus approaching an intersection. When there is such a potential, a radio announcement by the units approaching the intersection will help reduce the possibility of an accident.

Safe driving also requires that upon return from a call, the driver/operator perform routine checks of the truck. Such things as the tires, fuel level, oil level, radiator, water tank level, and general overall appearance need to be checked. This is in addition to the requirement that a complete check of the truck's condition be performed daily.

When driving on snow or ice, special precautions must be undertaken. These include:

1. Driving slowly and getting the "feel" of the road,
2. Using chains on snow or ice,
3. Ensuring that the windshield is kept clear and that the defroster is working efficiently,
4. Downshifting or applying the brake carefully to prevent the truck from going into a skid, and
5. Ensuring that there is sufficient space between the pumper and the vehicles in front to allow for stopping.

Driving the pumper

It is impossible to learn to drive a fire truck by reading a book. However, certain procedures can be described and then put into practice. The key is to actually drive the truck and learn under the supervision of a safety-conscious instructor. The basic procedures are:

1. When getting in the cab, adjust the seat so that you can comfortably reach the brake, clutch, and accelerator.
2. Adjust both left and right mirrors. Remember, there is no rear-view mirror on a pumper. You must get used to using the right-hand mirror.
3. Depress the clutch and make sure the gear selector is in neutral. Trucks with automatic transmission must have the selector in neutral.
4. Start the engine by turning on the electrical selector switch and pressing the starter button. Ensure that all gages are reading correctly.
5. Release the emergency brake(s) and place foot on foot brake.
6. Shift the gear selector into first gear, ease out on the clutch slowly while simultaneously pressing on the accelerator. The clutch must be engaged slowly to prevent putting a sudden load on the engine, transmission, and clutch. However, the clutch has to be let out fast enough to prevent stalling the engine. The necessary coordination can be learned with practice, with the final test being able to accelerate smoothly after stopping on a hill.
7. When the engine reaches the maximum rpm for first gear, the clutch should be depressed and the gear selector shifted to second gear. Up-shifting continues through the gears as the truck accelerates until the highest gear is reached. In city driving, the truck may not attain the top speed necessary to reach the highest gear.

8. One of the advantages of a manual transmission in a pumper is the ability to down-shift as the truck slows down. This helps to maintain control and allows the engine to assist the brakes in slowing down the truck. To down-shift, the driver depresses the clutch and shifts the gear selector to the next lower gear. This will provide an increase in engine rpm. As the slowdown continues, the truck is shifted to the next lower gear when engine rpm reaches a certain minimum value. Remember to keep the engine in gear when decelerating. Do not coast. The engine tachometer in the cab is the indicator to use for the correct time to shift. One other advantage of down-shifting is that the truck is ready to accelerate as conditions change.

9. Backing up a pumper is a very exacting procedure. Because there is no view directly to the rear of the pumper, an individual should be posted in the back during any reverse movement. In addition, a continual view of the left and right mirrors needs to be maintained. It is sometimes easier and quicker to go forward and a block or two out of the way to avoid the problem of backing up.

10. When stopping the pumper at the scene, make sure the emergency brake is set. Put wheel chocks under the wheels to prevent the truck from rolling.

11. When returning to the station, make the driver's maintenance checks to ensure that the truck is ready to respond again.

DRIVER TRAINING

Before beginning driver training, the fire department should administer a written test covering such items as:

Local and state laws
Area and street locations
Location of equipment on the apparatus
Departmental rules and regulations pertaining to driving.

The driver/operator should receive initial skills level training in a safe, large, open area that can be used to simulate driving conditions.

"The fire apparatus driver/operator, given a fire department pumper, shall demonstrate the following driving tests:
 "(a) Serpentine
 "(b) Alley dock
 "(c) Opposite alley pull in
 "(d) Diminishing clearance
 "(e) Straight line
 "(f) Turn around."*

Figure 1. Serpentine

Serpentine — The serpentine (figure 1) consists of a line of markers that are spaced 34 feet apart. The driver pulls through the course, going around

---

*Paragraph 3-1.7. Reprinted with permission from NFPA 1901-1982, Standard for Fire Apparatus Driver/Operator Professional Qualifications, Copyright© 1982, National Fire Protection Association, Quincy, Massachusetts 02269. This reprinted material is not the complete and official position of the NFPA on the referenced subject, which is represented only by the standard in its entirety.

the markers and then backs the apparatus through the markers. The entire course, forward and backward, must be traveled without stopping and without hitting any of the markers.

This driving procedure simulates the right control necessary for getting through and around traffic and fixed obstacles.

Alley dock — The alley dock (figure 2) consists of a marked area, just slightly larger than the truck. The driver has to pull up parallel to the opening and then back into it. This driving procedure simulates backing into the fire station or a narrow alley.

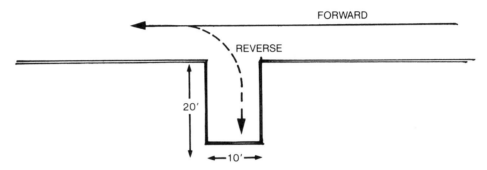

Figure 2. Alley Dock

Opposite alley pull in — In the opposite alley pull in, a 10-foot-wide area is offset by one apparatus length so that the driver has to move the truck to the right (figure 3). This driving procedure tests the driver's ability to manuever the vehicle while maintaining speed and keeping the apparatus under control.

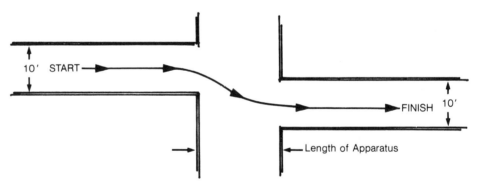

Figure 3. Opposite Alley Pull In

Diminishing clearance — In the diminishing clearance setup, the markers start at 9 feet 6 inches and diminish to 8 feet 2 inches over a 75-foot distance (figure 4). The driver must remain in the cab of the truck, using only

Figure 4. Diminishing Clearance

the front window to watch the course, and must stop within 6 inches of the finish. This driving procedure tests the driver's ability to control the vehicle and judge the clearance on both sides.

Straight line — In the straight line test, the markers are set to determine if the driver keeps the vehicle on a straight line (figure 5). In this test, the markers

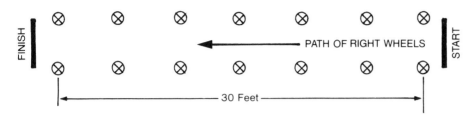

Figure 5. Straight Line

are set up for 30 feet and separated by 2 feet 6 inches. The right wheel of the truck is placed between the markers and the distance must be traveled without knocking down any of the markers.

Turn around — In the turn around exercise (figure 6) the driver's ability to make a three-point turn to reverse direction is tested. Both a left and right turn around are tested because of the difference in maneuvering required for each.

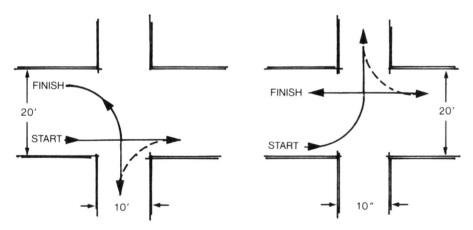

Figure 6. Turn Around

Appendix A

Understanding Fluids

Before defining a fluid, an explanation of the characteristics that a fluid possesses will aid in understanding the scientific principles involved. All matter can exist in any one of three states: solid, liquid, or gas, which are composed of tiny particles called *molecules*.

In a solid, the molecules are spaced very close together, and by their nature they try to hold onto each other. Now when a solid is heated, the molecules vibrate. As more heat is applied, the molecules vibrate more and more rapidly and try to break away from each other. At some point, the molecules do break away and the form of the solid changes to either a liquid or gas. The stronger the attachment of the molecules of a solid, the higher heat that is necessary to change the form.

Definition of a fluid

A fluid is defined as a substance that yields to the slightest force and recovers its previous state when the force is removed. The molecules of a fluid are not closely bound and cannot sustain a sideways force. This is the basic distinguishing factor between a fluid and a solid. Fluids can be divided into two subclasses — gases and liquids.

An understanding of fluid behavior is necessary because the firefighter must be able to anticipate actions of fluids. From flammable liquid spills to toxic gases, an understanding of the properties of fluids will aid the pump operator.

In order to understand what causes fluids to move, a discussion of the various types of forces is necessary. The three major forces that can act on any body are shear, tensile, and compressive.

A *shear force* (figure 1) is a sideways force which will cause the fluid to move. A *tensile force* (figure 2) is an attempt to pull apart the fluid and does not cause much movement. The only motion of a tensile force is caused by the ability of the particular molecules to stick together. A *compressive force* (figure 3) is the attempt to push the fluids together. This will cause the best movement of the fluid and, of the three forces, permits the best control.

The compressive forces used in the fire service, when referenced to a given area, are known as *pressure*. Pressure denotes a force per unit area. In the American system of measurement, force is measured in pounds and area is usually measured in square inches or square feet. In the fire service, pressure in force per unit area is expressed as pounds per square inch and abbreviated as psi.

UNDERSTANDING FLUIDS

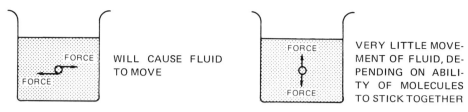

Figure 1. Shear Force

Figure 2. Tensile Force

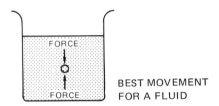

Figure 3. Compressive Force

Both liquids and gases are fluids. The properties of fluids (figure 4) can be generalized as:

Liquids

Almost incompressible. (For purposes of this text, a liquid will be considered incompressible because it takes a pressure of 30,000 psi to cause a 1 percent reduction in a given volume of water.)

Assumes the shape of the container. (Whether in a glass, lake or pool, a liquid always conforms to the shape of the vessel.)

Occupies a definite volume, independent of the shape of the vessel. (Whether it's confined to a tall, thin glass or a long, flat plate, the liquid will have the same volume.)

Capable of flowing. (The fact that a liquid is not compressible permits it to flow easily.)

Gases

Compressible. (A 40-cubic-foot volume of air is compressed into a small cylinder for breathing apparatus.)

Assumes the shape of the container. (A gas will expand until it fills the vessel holding it, whether it be a jar or a room.)

Will expand to occupy the volume of the container. (If a bottle of gas were released in a room, the gas would continue to expand until it took the shape of the room.)

Capable of flowing under certain circumstances. (Air will flow in a ventilating system or around the wing of an airplane. Under these special conditions, the gas acts as if it is incompressible.)

The prime fluid that pump operators are concerned with is water. Water has some properties that make its use on the fireground beneficial.

Properties of water

There are two different conditions that a firefighter must consider when operating a pump. The first is when the fluid is at rest or under static conditions. (The study of fluids at rest is called hydrostatics.) The second possibility is

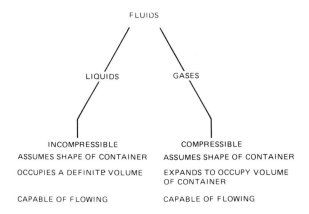

Figure 4. Properties of fluids

that the fluid is in motion or flowing (the study of this type of motion is called hydrokinetics).

First, consider the physical properties of water. Water is composed of molecules, with each molecule made up of two atoms of hydrogen and one atom of oxygen (figure 5). The chemical symbol for this is written as H_2O. In their natural state, hydrogen and oxygen are gases, but when combined to form a molecule of water, they become a completely new substance. This new substance is a liquid between 32°F (freezing point) and 212°F (the boiling point at normal atmospheric pressure). In addition, once hydrogen and oxygen form water, it is very difficult to separate back to their original elements.

Under static conditions, water follows five basic laws:

1. Fluid pressure is perpendicular to the surface on which it works (figure 6A).

This means that when water is confined by a barrier, it causes pressure which acts at 90 degrees to the barrier. What would happen if this were not true? As shown in figure 6B, assume that the force acts at some other angle than 90 degrees. Under this condition, the force would cause the water to move and a static condition would not be present. The only time there will be a static condition is when the force is perpendicular to the barrier.

2. At any point within the water, the pressure is the same amount in all directions (figure 6C).

Within the water, at any point the pressure exerted by water molecule "X" on the surrounding molecules is the same in all directions. It is important to distinguish that this does not say the pressure at point "X" is equal to the pressure at point "Y."

3. Pressure from the outside, when applied to water in a confined area will be distributed to all parts of the area without decreasing in value (figure 6D).

Because water is incompressible, an external force of 100 psi applied to the water will cause the pressure throughout the vessel to increase by 100 psi.

4. The pressure exerted by water at any point in the open vessel is dependent upon its depth (figure 6E).

5. The pressure exerted by water does not depend upon the shape of the vessel (figure 6F).

Pressure for the fire service is measured in pounds per square inch. Because everything is referenced to this particular area, the shape of the vessel will not change the pressure reading. Only the height of the water above the gage will change the pressure. Gages Q, R and S all read in psi, and because the height is the same, all gages will read equal pressures. Note that the size and capacity of the containers is not mentioned, yet the pressure will be the same.

UNDERSTANDING FLUIDS

Figure 5. Chemistry of water

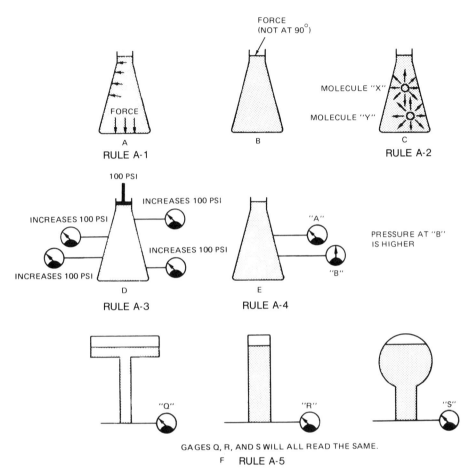

Figure 6. Properties of water

The definition of pressure and the distinction between pressure and force have been discussed. In addition, the basic laws of pressure as applied to water under static conditions have been covered.

255

Appendix B

Water Movement

Flow or movement of water can occur in any of three directions at the same time. The movement of a fluid in three directions is called *three-dimensional flow*. For the applications and uses in the fire service, it is sufficiently accurate to think in terms of flow or movement only occurring in one direction. This is called *one-dimensional flow* (figure 1).

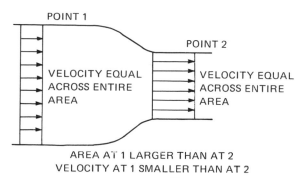

Figure 1. One-dimensional flow

As the water passes through hose, look at the flow in two places, indicated at points 1 and 2 on figure 1. (Keep in mind that one-dimensional flow is an approximation and cannot be extended to more complicated flows in other fluids.)

If you could stand at one point with a counter and a stop watch, you would be able to count how many cubic feet of water passed by every minute. The amount of water would be determined by how fast the water was flowing and how big was the area of the hose. From this information the formula for flow can be written:

$$Q = AV, \text{ where}$$
$$Q = \text{quantity of flow}$$
$$A = \text{area}$$
$$V = \text{velocity of flow}$$

It is extremely important, when using this formula, to use the proper units. If the area is measured in square inches and the velocity in inches per minute, then flow would be cubic inches per minute. This can be stated as:

$$Q = A \text{ (inches}^2\text{)} \times V \left(\frac{\text{inches}}{\text{minute}}\right)$$

$$Q = \frac{\text{inches}^3}{\text{minute}}$$

WATER MOVEMENT

However, in the fire service, the common measure for flow is gallons per minute. So this formula must now be converted so that Q is expressed in gpm:

$$Q = A\;(\text{in}^2) \times V\left(\frac{\text{in}}{\text{min}}\right)$$

1 gallon contains 231 cubic inches of water

$$Q = \frac{A\;(\text{in}^2) \times V\left(\frac{\text{in}}{\text{min}}\right)}{231\;\frac{\text{in}^3}{\text{gal}}}$$

Using the division of fractions rule of inverting and multiplying

$$Q = \frac{A\;(\text{in}^2) \times V\;(\text{in})}{(\text{min})} \times \frac{1\;(\text{gal})}{231\;(\text{in}^3)}$$

The in³ cancels with the in² × in, leaving

$$Q = \frac{\text{gallons}}{\text{minute}}$$

Example: What is the flow in gallons per minute in a 3-inch hose line, if the water is moving past a point at the rate of 102 feet per minute?

Step 1. Determine the area:

$$\text{Area of a circle} = \pi r^2$$
$$r = 1\tfrac{1}{2} \text{ inches}$$
$$\pi = 3.14$$
$$A = 3.14\;(1\tfrac{1}{2})^2$$
$$A = 3.14\;(2.25)$$
$$A = 7.06\;\text{in}^2$$

Step 2. Determine the velocity in correct units:

$$V = 102\;\frac{\text{ft}}{\text{min}} \times 12\;\frac{\text{in}}{\text{ft}}$$
$$V = 102\;(12)\;\frac{\text{in}}{\text{min}}$$
$$V = 1224\;\frac{\text{in}}{\text{min}}$$

Step 3. Solve the equation:

$$Q = AV$$
$$Q = \frac{7.06\;\text{in}^2 \times 1224\;\frac{\text{in}}{\text{min}}}{231\;\frac{\text{in}^3}{\text{gal}}}$$
$$Q = \frac{8641\;\text{gal}}{231\;\text{min}}$$
$$Q = 37.41\;\text{gpm}$$

Conservation of matter

Now, just from intuition, it can be said that if 200 gallons per minute were put in one end of a hose line, the other end would flow 200 gpm when using a noncompressible fluid. Notice that nothing was said about the size of the hose. What goes in one end must come out the other end at the same rate. In physics this rule is stated as: Matter can be neither created nor destroyed.

This rule means that at any point in the system the flow rate must be equal. The flow rates at points 1 and 2 of figure 1 can now be written as

$$Q_1 = A_1 V_1$$
$$Q_2 = A_2 V_2 \text{ and since the flows are equal}$$
$$Q_1 = Q_2 \text{ and}$$
$$A_1 V_1 = A_2 V_2$$

This means that the larger the area gets the smaller the velocity will be.

Example: using figure 1, the area at point 1 is 4 square inches and the velocity is 220 feet per second. What is the velocity at point 2 in feet per second, if the area is 3 square inches?

Step 1. Determine the area:

$$A_1 = 4 \text{ in}^2$$
$$A_2 = 3 \text{ in}^2$$

Step 2. Determine the velocity in correct units:

$$V_1 = 220 \frac{ft}{sec} \times 12 \frac{in}{ft}$$

$$V_1 = 2640 \frac{in}{sec}$$

$$V_2 = \text{unknown}$$

Step 3. Solve the equation:

$$A_1 V_1 = A_2 V_2$$

$$4 \text{ in}^2 (2640) \frac{in}{sec} = 3 \text{ in}^2 (V_2)$$

$$\frac{4 (2640)}{3} \frac{in}{sec} = V_2$$

$$\frac{10560}{3} \frac{in}{sec} = V_2$$

$$3520 \frac{in}{sec} = V_2$$

Step 4. Convert to correct units:

$$\frac{3520 \frac{in}{sec}}{12 \frac{in}{ft}} = V_2$$

$$V_2 = 293.33 \frac{ft}{sec}$$

Note that as the area decreased at point 2 to 3 inches the velocity increased to 293.33 ft/sec.

Bernoulli's Equation

In order to continue the discussion of fluids in motion, certain further definitions are necessary. The energy possessed by a fluid can be defined as its capacity to do work. This ability to do work can be classified into groups such as electrical, chemical, atomic, thermal, and mechanical. For the field of pump operation the main group is mechanical energy.

Mechanical energy can be divided into two major areas: potential energy and kinetic energy. *Potential energy* of a fluid is that energy which the fluid has stored, ready for use. One example would be an elevated water tank in which the water has potential energy to do work because it is elevated above the ground. The *kinetic energy* of a fluid is its energy due to motion. For example, water flowing from a nozzle possesses kinetic energy because the flowing water is able to do work (turn a wheel or knock down a wall).

Potential energy can be stored in one of two ways. It can receive its energy due to elevation (head) or due to pressure under static conditions. Since potential energy can also be a combination of these two, a generalized equation can be written:

total potential energy = pressure head + elevation head

It is important to keep the units of this equation consistent. Each value must be expressed in terms of equivalent head, which in the fire service is usually feet.

The rule which states that matter cannot be created nor destroyed can now be extended to include energy. Energy can be changed from one form to another, but the total available energy must remain constant. Because of this the total energy possessed by a fluid can be written as:

$$\text{Total energy} = \underbrace{\text{pressure head} + \text{elevation head}}_{\text{potential energy}} + \text{kinetic energy}$$

In keeping with the requirements of expressing the energy in terms of head, kinetic energy will be called velocity head.

Figure 2 shows how a hydraulic system receives energy. With no water flowing, what is the gage reading at the hydrant? Potential energy at the hydrant is due to the elevation head plus the static pressure head created by the pump. With no water flowing, the velocity head equals zero. The equation can now be written:

total energy = pressure head + elevation head + velocity head

pressure head (PH) = pressure in psi × 2.31 ft/psi

PH = 100 psi × 2.31 ft/psi

PH = 231 ft

Elevation head (EH) = 200 ft

Velocity head (VH) = 0 (static conditions)

total energy = 231 + 200 + 0 = 431 ft

Pressure at hydrant = P = h × .434

P = 431 × .434 = 187 psi

PUMP OPERATORS HANDBOOK

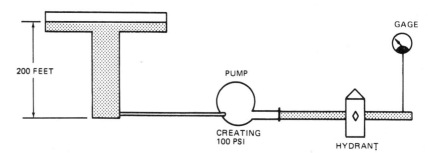

Figure 2. Total energy under static conditions

Now, energy cannot be created nor destroyed. Therefore, at any one point in the system the total energy must be equal to the total energy at any other point in the system. At points 1 and 2 in figure 3, the energy equation can be written as:

$$\text{pressure head}_1 + \text{elevation head}_1 + \text{velocity head}_1 =$$
$$\text{pressure head}_2 + \text{elevation head}_2 + \text{velocity head}_2$$

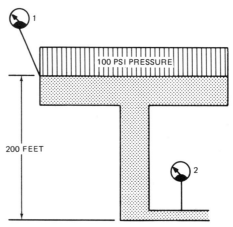

Figure 3. Bernoulli's Equation

This statement about the equality of energy at any point in a fluid system is known as Bernoulli's Equation.

Using figure 3, what is the velocity head at point 2?

At point 1, $PH_1 = 100 \times 2.31 = 231$ ft

$EH_1 = 200$ ft

$VH_1 = 0$ (even though water is moving, the area is so large, a velocity of 0 can be assumed.)

At point 2, $PH_2 = 0$ (discharge open to the atmosphere so there is no pressure head.)

$EH_2 = 0$

$VH_2 = $ unknown

$$PH_1 + EH_1 + VH_1 = PH_2 + EH_2 + VH_2$$
$$231 + 200 + 0 = 0 + 0 + VH_2$$
$$VH_2 = 431 \text{ ft}$$

Within the fire service, velocity head by itself will not be useful to the pump operator. A more beneficial value is the velocity reached by the fluid.

Velocity formula

The velocity attained by a falling object is determined by the height through which it will fall due to gravity pulling it down. Since gravity is a constant value, the farther the object must drop the longer it will be able to increase its velocity. This means that the longer it drops the faster it will be going (higher velocity) when it reaches the bottom. The elements of time and gravity can be related to velocity with the formula:

$$v = gt, \text{ where}$$
$$v = \text{velocity in ft/sec}$$
$$g = \text{gravity which is } 32.2 \text{ ft/sec}^2$$
$$t = \text{time in seconds}$$

So, if it takes 3 seconds for an article to fall, its velocity will be:

$$v = 32.2 \frac{ft}{sec^2} \times 3 \text{ sec}$$
$$v = 96.6 \frac{ft}{sec}$$

Note that the size, shape and weight do not enter into the picture. The velocity obtained is based solely on the action of gravity.

The velocity of a falling object does not remain in a constant. At the end of one second the velocity is 32.2 ft/sec and at the end of two seconds it is 64.4 ft/sec. The average velocity of the object can be obtained by taking the final velocity and dividing by 2. This gives the formula:

$$\text{average velocity} = \frac{v}{2} = \frac{gt}{2}$$

If the average velocity was 100 ft/sec, the height through which the object fell can be determined by multiplying by the time it took to fall. In this example, if it took 10 seconds the object fell

$$100 \frac{ft}{sec} \times 10 \text{ sec} = 1000 \text{ ft}$$

Using this as a base, then multiplying the average velocity by time to yield the height, h, which is equal to velocity head produces the formula

$$h = \frac{v}{2}(t)$$
$$= \frac{gt}{2}(t)$$
$$= \frac{gt^2}{2}$$

Using the formula $v = gt$ and solving for t yields

$$t = \frac{v}{g}$$

Substituting yields

$$h = \frac{g}{2}\left(\frac{v}{g}\right)^2$$

$$h = \frac{\cancel{g}}{2} \times \frac{v}{\cancel{g}} \times \frac{v}{g}$$

$$h = \frac{v^2}{2g}$$

Solving for v gives

$$v^2 = 2gh$$
$$v = \sqrt{2gh}$$

Since g = 32/2 ft/sec, this formula can be written as

$$v = \sqrt{2 \times 32.2 \times h}$$
$$v = \sqrt{64.4\,h}$$
$$v = 8.02\sqrt{h}$$

The velocity head or height is related to pressure in psi through the formula

$$h = 2.31\,P$$

Substituting in the velocity formula yields

$$v = 8.02\sqrt{2.31P}$$
$$v = 8.02 \times 1.52\sqrt{P}$$
$$v = 12.19\sqrt{P}$$

Now, the velocity can be calculated when either the pressure or the elevation head are known.

Example: With a nozzle pressure of 80 psi, what is the velocity of discharge?

Step 1. Select the formula to use:

$$v = 12.19\sqrt{P}$$

Step 2. Determine the formula values:

$$P = 80 \text{ psi}$$

Step 3. Solve the equation

$$v = 12.19\sqrt{80}$$
$$v = 12.19\,(8.94)$$
$$v = 109 \text{ ft/sec}$$

Example: If a hole is made in a water tank 94 feet below the surface of the water, what would be the velocity of the water as it is discharged from the hole?

Step 1. Select the formula to use:

$$v = 8.02 \sqrt{h}$$

Step 2. Determine the formula values:

$$h = 94 \text{ ft}$$

Step 3. Solve the equation:

$$v = 8.02 \sqrt{94}$$
$$v = 8.02 \,(9.70)$$
$$v = 77.8 \text{ ft/sec}$$

The form of Bernoulli's Equation for water can now be rewritten to reflect the formulas just developed.

$$PH_1 + EH_1 + VH_1 = PH_2 + EH_2 + VH_2$$

$$PH = \frac{P}{.434}$$

$$EH = z \text{ (elevation)}$$

$$VH = \frac{v^2}{2g}$$

$$\frac{P_1}{.434} + z_1 + \frac{v_1^2}{2g} = \frac{P_2}{.434} + z_2 + \frac{v_2^2}{2g}$$

This is usually the way Bernoulli's Equation is expressed in hydraulic textbooks. It is just another way of saying that the sum of the energies within a hydraulic system are the same.

Measurement devices

There are two major devices for making measurements of flow and velocity: the venturi meter and the pitot gage. Both of these devices use the formulas developed in this unit to make measurements.

The venturi meter, figure 4, uses Bernoulli's Equation

$$\frac{P_1}{.434} + z_1 + \frac{v_1^2}{2g} = \frac{P_2}{.434} + z_2 + \frac{v_2^2}{2g}$$

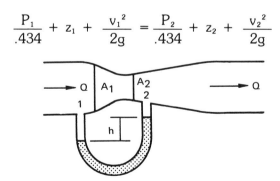

Figure 4. Venturi meter

If the meter is level, there is no elevation and z_1 and z_2 equal zero. The equation then reduces to

$$\frac{P_1}{.434} + \frac{v_1^2}{2g} = \frac{P_2}{.434} + \frac{v_2^2}{2g}$$

Now, compare area at points 1 and 2 of figure 4. The area at point 2 is smaller so the velocity will be higher. In the formula an increase in velocity at point 2 will cause a decrease in pressure. The lower pressure at point 2 will cause the liquid (usually mercury) in the tube to deflect. The difference in height, h, is calibrated so that velocity and flow can be read from a table.

A pitot gage, figure 5, is also used to measure velocity head at the discharge

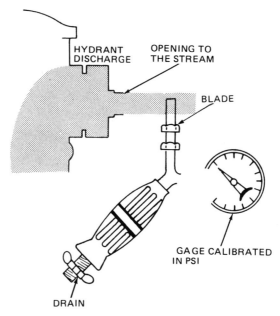

NOTE: THE OPENING IN THE BLADE MUST BE PLACED IN THE CENTER OF THE STREAM BECAUSE THE VELOCITY HEAD IS NOT QUITE EQUAL ACROSS THE ENTIRE DISCHARGE OPENING.

Figure 5. Pitot gage

of a nozzle. From the reading of velocity head, h, velocity can be calculated from the formula

$$v = 2gh$$

Then, flow can be calculated using the formula

$$Q = AV$$

In actual practice, a gage is connected to the pitot tube so that readings directly in psi can be made. Then, using the size of the nozzle and the discharge pressure, the flow can be read from a table.

Appendix C

Quantity of Water Flowing

The first step in calculating proper engine pressure is to determine the amount of water flowing. On the fireground this can be arrived at by means of educated guesses, but the pump operator must understand how these figures are determined. He will thus be able to handle unique situations, new appliances, and new apparatus without difficulty.

It has been shown that discharge through a hose line was equal to the area times the velocity.

$$Q = AV$$

Due to gravity, the velocity of flow was shown to be

$$V = 12.14 \sqrt{P} \; \frac{ft}{sec}$$

Substituting yields

$$Q = A \; ft^2 \times 12.14 \sqrt{P} \; \frac{ft}{sec} = \frac{ft^3}{sec}$$

Since the common measure for flow, Q, in the fire service is gallons per minute and area of the nozzle is expressed in inches, this formula must be converted for pump operator use.

Area of nozzle discharge $A = \pi r^2$

The radius, r, of a circle is ½ of the diameter, d.

$$r = \frac{d}{2}$$

Area of a nozzle $A = \pi \times \frac{d}{2} \times \frac{d}{2}$

$$A = \frac{\pi d^2}{4}$$

In the formula A is measured in feet, so the formula must be converted so that d can be expressed in inches.

$$A = \frac{\pi \times d^2 \, \cancel{in^2}}{4} \times \frac{1 ft^2}{144 \, \cancel{in^2}}$$

$$A = \frac{\pi d^2}{576} ft^2 \text{ with d measured in inches.}$$

Velocity in feet per second must be converted so that Q will be measured in gpm.

$$V = 12.14 \sqrt{P} \, \frac{ft}{sec}$$

$$V = 12.14 \sqrt{P} \, \frac{ft}{sec} \times \frac{60 \, sec}{1 \, min}$$

$$V = 728.4 \sqrt{P} \, \frac{ft}{min}$$

There are 7.481 gallons per cubic foot.

$$V = 728.4 \sqrt{P} \, \frac{ft}{min} \times 7.481 \, \frac{gal}{ft^3}$$

$$V = 5449.16 \sqrt{P} \, \frac{gal}{min \times ft^2}$$

Substituting the values for A and V in the original equation yields

$$Q = \frac{\pi d^2 \, ft^2}{576} \times 5449.16 \sqrt{P} \, \frac{gal}{min \times \cancel{ft^2}}$$

$$Q = \frac{3.1416 \, d^2}{576} \times 5449.16 \sqrt{P} \, \frac{gal}{min}$$

$$Q = \frac{17,119.08}{576} d^2 \sqrt{P}$$

$$Q = 29.72 \, d^2 \sqrt{P} \, \frac{gal}{min}$$

This formula is known as the Freeman Formula. The American Insurance Association uses a slight variation of this formula

$$Q = 29.83 \, d^2 \sqrt{P} \, \frac{gal}{min}$$

Both formulas are very close in value, with a variation of less than 1 percent. Calculations within the text use Freeman's Formula.

One of the problems with both of these formulas is that the value is purely theoretical because friction loss within the device is not considered. To correct the equation, a coefficient of discharge, c, is included in the equation. Freeman's equation can then be written as

$$Q = 29.7 \times d^2 \times \sqrt{P} \times c$$

The value of the coefficient of discharge will vary with the opening through which the water will flow. The 2½-inch discharge opening of the fire hydrant

QUANTITY OF WATER FLOWING

can have a value of c which varies between 0.7 and 0.9. Smooth-bore nozzles have a c value ranging from 0.96 to 0.99. Since the values of 0.96 to 0.99 are all close to 1.0, the coefficient is sometimes deleted from the calculations for smooth-bore nozzles.

Examples: How much water will be delivered from a 1¼-inch smooth-bore nozzle if the nozzle pressure is 50 psi?

Step 1. Write the formula:

$$Q = 29.7 \times d^2 \times \sqrt{P} \times c$$

Step 2. Determine the formula value:

$$d = 1.25 \text{ inches}$$
$$P = 50 \text{ psi}$$
$$c = 1$$

Step 3. Solve the equation:

$$Q = 29.7 \times (1.25)^2 \times \sqrt{50} \times 1$$
$$Q = 29.7 \times 1.56 \times 7.07 \times 1$$
$$Q = 327.55 \text{ gpm}$$

Example: How much water will be delivered from a 2½-inch discharge of a hydrant with a coefficient of discharge of 0.85 and a discharge pressure of 8 psi?

Step 1. Write the formula:

$$Q = 29.7 \times d^2 \times \sqrt{P} \times c$$

Step 2. Determine the formula values:

$$d = 2.5 \text{ inches}$$
$$P = 8 \text{ psi}$$
$$c = 0.85$$

Step 3. Solve the equation:

$$Q = 29.7 \times (2.5)^2 \times \sqrt{8} \times 0.85$$
$$Q = 29.7 \times 6.25 \times 2.83 \times 0.85$$
$$Q = 446.5 \text{ gpm}$$

Solving friction loss using K factors for 2½-inch hose

For flow in a straight-bore nozzle with a diameter of 1 inch or greater, the formula is:

$$FL = 1.1 \times K \times NP \times L, \text{ where}$$
$$FL = \text{friction loss in 2½-inch hose}$$

$$K = \frac{d^4}{10} \text{ for 2½-inch hose only}$$

d = nozzle diameter in inches

NP = nozzle pressure in psi

L = number of 50-foot lengths of hose

Example: What is the friction loss in 400 feet of 2½-inch hose using a 1¼ inch tip at 42-psi nozzle pressure?

Step 1. Select the proper equation, when flow is known:

$$FL = 1.1 \times K \times NP \times L$$

Step 2. Determine the formula values:

$$K = \frac{d^4}{10}$$

$$K = \frac{(1¼)^4}{10}$$

$$K = \frac{1.25 \times 1.25 \times 1.25 \times 1.25}{10}$$

$$K = \frac{2.44}{10}$$

$$K = .244$$

$$NP = 42 \text{ psi}$$

$$L = \frac{400}{50} = 8$$

Step 3. Solve the equation:

$$FL = 1.1 \times .244 \times 42 \times 8$$

$$FL = 90 \text{ psi for 400 ft}$$

Friction loss calculations using K for small flows

If the nozzle diameter is less than 1 inch, use the formula

$$FL = 1.0 \times K \times NP \times L, \text{ where}$$

K = value determined from table

NP = nozzle pressure

L = number of 50-foot lengths of hose

Because the values of K are dependent upon two constants, nozzle diameter and hose size, they can be calculated in advance. Table 1 lists the values of K.

Example: What is the friction loss in 200 feet of 1½-inch hose using a ¾-inch nozzle at 56-psi nozzle pressure?

Step 1. Select the proper equation for unknown flow with a nozzle diameter less than 1 inch:

$$FL = 1.0 \times K \times NP \times L$$

Step 2. Determine the formula values:

$$K = .455 \text{ (table 1)}$$
$$NP = 56 \text{ psi}$$
$$L = \frac{200}{50} = 4$$

Step 3. Solve the equation:

$$FL = 1.0 \times .455 \times 56 \times 4$$
$$FL = 102 \text{ psi per 200 ft}$$

TABLE 1. Values of K

Nozzle diameter	1½-inch line	2½-inch line	3-inch line	Dual 2½-inch lines	Dual 3-inch lines
	use small diameter formula				
1/4	.0084				
3/8	.036				
1/2	.104	.0115			
5/8	.234	.023			
3/4	.455	.042			
7/8		.068			
1		.105	.038	.025	
1 1/8		.167	.062	.043	
1 1/4		.248	.092	.066	.023
1 3/8		.341	.137	.096	.034
1 1/2		.505	.192	.135	.051
1 5/8		.680	.266	.184	.068
1 3/4		.907	.351	.242	.093
2		1.550	.605	.418	.157

Engine pressure calculations using K factors

One of the formulas developed for friction loss was

$$FL = 1.1 \times K \times NP \times L$$

Substituting this in the engine pressure formula

$$EP = NP + FL \pm E$$
$$EP = NP + 1.1 K \times NP \times L \pm E$$

If level ground is assumed, then E = 0 and the formula becomes

$$EP = NP + 1.1 K \times NP \times L$$
$$EP = NP [1 + 1.1 K(L)]$$

With a very slight introduction of error, the expression within the bracket can be written as

$$[1.1 + K (L)]$$

Then,

$$EP = NP [1.1 + K(L)]$$

This formula is known as the Underwriters formula and is used for calculating engine pressure when flow is not known.

As with the friction loss formula, engine pressure calculation for hose with a nozzle smaller than 1 inch and 2½-inch or less hose must use the formula

$$EP = NP [1 + K(L)]$$

Example: What engine pressure is necessary to deliver 50-psi nozzle pressure with a 1-1/8-inch tip through a 300-foot length of 2½-inch hose?

Step 1. Select the proper equation, with tip size larger than 1 inch.

$$EP = NP[1.1 + K(L)]$$

Step 2. Determine the formula values:

$$NP = 50 \text{ psi}$$
$$K = .167 \text{ (table 1)}$$
$$L = \frac{300}{50} = 6$$

Step 3. Solve the equation:

$$EP = 50 [1.1 + .167 (6)]$$
$$EP = 50 [1.1 + 1.0]$$
$$EP = 50 [2.1]$$
$$EP = 105 \text{ psi}$$

Appendix D

Equations

Average velocity	$\dfrac{V}{2} = \dfrac{gt}{2}$
Back pressure	$P = .434\,h$
Bernoulli's Equation	$PH_1 + EH_1 + VH_1 = PH_2 + EH_2 + VH_2$
	$\dfrac{P_1}{.434} + z_1 + \dfrac{v_1^2}{2g} = \dfrac{P_2}{.434} + z_2 + \dfrac{v_2^2}{2g}$
Engine pressure	$EP = NP + FL \pm E$ (hose line)
	$EP = NP\,[1.1 + K(L)]$ (larger than 1 inch)
	$EP = NP\,[1 + K(L)]$ (smaller than 1 inch)
	$EP = FL + RP \pm E$ (relay)
Flow	$Q = AV$
	$Q = 29.7 \times d^2 \times \sqrt{P} \times c$ (Freeman)
	$Q = 29.83 \times d^2 \times \sqrt{P} \times c$ (AIA)
	$Q = \frac{1}{2}P + 15$ (sprinkler head)
	$Q_1 + Q_2 = A_1V_1 = A_2V_2$
Friction loss (100 feet of 2½-inch hose flowing less than 100 gpm	$FL = 2Q^2 + \frac{1}{2}Q$
	$FL = 1.0\,K \times NP \times L$ (smaller tip than 1″)
Friction loss (100 ft of 2½″ hose flowing more than 100 gpm	$FL = 2Q + Q$
	$FL = 1.1\,K \times NP \times L$
Net pump pressure	$NPP = psig + \dfrac{\text{lift (ft)}}{2.3} + \dfrac{\text{suction loss (ft)}}{2.3}$
Nozzle reaction	$NR = 1.57 \times d^2 \times P$ (straight tip)
	$NR = .0505 \times Q \times \sqrt{P}$ (fog nozzle)
Velocity	$v = gt$
	$v = \sqrt{2gh}$
	$v = 8.02\,\sqrt{h}$
	$v = 12.19\,\sqrt{P}$
Velocity head	$h = 2.31P$

Appendix E

Definitions

Absolute pressure: True pressure, which equals the sum of atmospheric and gage pressures (psia).
Accelerator: Bleeds air from a dry pipe sprinkler system rapidly.
Air pressure gage: Shows the pressure available in the brake lines.
Algebraic expression: A group of symbols which represents a number.
Ammeter: Shows how much current is flowing into or out of the battery.
Associative rule of addition: Addition can be performed in any sequence.
Associative rule of multiplication: Multiplication can be performed in any sequence.
Atmospheric pressure: Pressure caused by the elevation of air above the earth.
Auxiliary cooling valve: Permits water from the pump to cool the radiator water through a heat exchanger.
Back pressure: Pressure caused by the elevation of water.
Baffles: Metal parts added to water tanks to prevent surging.
Barrel: On a hydrant, conducts water from the foot piece to the bonnet.
Bearings: On a pump, provide support and alignment for the impeller shaft.
Bonnet: On a hydrant, provides protection to the unit and contains the mechanism for turning the valve.
Bourdon tube: Hollow curved tube which activates a pressure gage.
Capacity: See parallel.
Cavitation: Caused by the pump attempting to deliver more water than is being supplied. This causes the formation of water vapor and liquid water, under pressure, rushes in to fill the empty space. This results in a tremendous shock.
Centrifugal force: Force which tends to make rotating bodies move away from the center of rotation.
Centrifugal pump: A pump which uses a rapidly spinning disk to create the pressure for water movement.
Certification: Pumper test conducted by Underwriters' Laboratories to determine if the pumper can deliver its rated volume and pressures.
Check valve: Valve which prevents water from flowing in the wrong direction.
Clearance rings: See wear rings.
Clover leaf pump: A rotary pump using three gear teeth.
Common fraction: A fraction which has both terms expressed.
Commutative rule of addition: Order of the addition is reversible.

Commutative rule of multiplication: The order of multiplication is reversible.
Complex fraction: A fraction which contains a whole number in the numerator or denominator, or in both of them.
Compound gage: Gage which indicates both positive and negative pressure on the same gage.
Compressive force: Force which tends to push things together.
Control valve: Valve which regulates flow of water from water main to the sprinkler system.
Dead-end main: Main that is not cross connected to any other main.
Deluge system: Sprinkler system which delivers water to a large area all at once.
Denominator: The number of equal parts that the whole is divided into.
Direct proportion: Unlike quantities change in the same order in each ratio.
Distribution mains: Small-size pipe which feeds individual streets of the service area.
Distributive rule of multiplication: One number times the sum of two other numbers is equal to the first number times one of the other numbers plus the first number times the second of the other numbers a × (b × c) = (a × b) + (a × c).
Dividend: Number to be divided.
Divisor: Number to divide by.
Double-acting pump: Type of piston pump which discharges water while the piston moves in either direction.
Double suction impeller: Water enters on both sides of the impeller.
Drafting: Using water from a static source.
Drip valve: Prevents water from accumulating in a dry sprinkler system due to a leak.
Dry barrel hydrant: Barrel of hydrant drains automatically after hydrant is shut down.
Dry prime test: Provides information on the ability of the pumper to evacuate air and draft water.
Dry pipe sprinkler system: Air pressure fills the piping and holds back the water from entering until needed.
Dry pipe standpipe: Standpipe system in which the pumper is the source of supply.
Dry pipe valve: Keeps the water out of a dry pipe sprinkler system.
Dual pumping: Connecting two pumpers intake to intake so that the second pumper receives the excess water available from the hydrant.
Dynamometer: Device to measure force.
Energy: Capacity to do work.
Equation: A statement which expresses equality.
Exhaust primer: Primer which uses the venturi principle of fast-flowing exhaust gases to remove air from the pump.
Exhauster: Bleeds air from a dry-pipe sprinkler system rapidly.
Extremes: Outside terms of a proportion.
Eye: The part of the impeller where water enters.
Factors: Individual parts of the terms in mathematics.
Flap valve: Valve that controls the flow of water inside a multistage pump.
Flinger ring: Prevents water from continuing to travel the impeller shaft to the gears and ball bearings.
Fluids: Substances which yield to the slightest force and recover their previous state when the force is removed.
Foot piece: Inlet for water to a hydrant.

Formula: General expression which states a fact.
Fraction: Part of a whole number.
Frangible bulb: A type of sprinkler head which activates by expansion of a liquid as it heats up.
Front-mount pump: Pump mounted ahead of the engine on a front engine type of apparatus.
Fusible link: Type of sprinkler head which activates by melting at a specific temperature.
Gage pressure: Pressure read on a gage (psig).
Governor: Minimizes pressure changes by controlling engine speed.
Grid: Different size pipes which are connected together to make a water distribution system.
Head: Height to which a given pressure will elevate water.
Horsepower: Amount of work which can be produced by an engine.
Hydraulics: Study of fluids.
Hydrokinetics: Study of fluids in motion.
Hydrostatics: Study of fluids at rest.
Impeller: Part of the centrifugal pump which provides velocity to the water.
Improper fraction: Fraction whose numerator is larger than its denominator.
Indirect proportion: One ratio gets larger while the other ratio gets smaller.
Inverse proportion: See indirect proportion.
Kinetic energy: Energy due to motion.
Laminar flow: Water moving in straight lines.
Lift pump: Special type of piston pump which develops low pressure.
Line gage: Indicates pressure on individual hose lines.
Looped main: Cross-connected water main.
Lowest common denominator: Smallest denominator that a group of fractions has in common.
Means: Inside terms of a proportion.
Midship pump: Pump mounted behind the cab of the apparatus.
Minuend: Larger number of a subtraction.
Mixed number: Number which contains a whole number and a fraction.
Multiplicand: Number to be multiplied.
Multiplier: Number of times the multiplicand is to be multiplied.
Needle valve: Installed on gage to permit a steady reading, without vibration.
Negative pressure: Pressure below atmospheric.
Numerator: Number of equal parts of the whole which has been taken to make a fraction.
Numerical coefficient: Number in a term.
Odometer: Records distance traveled in miles and sometimes continues to function while pumping.
Oil pressure gage: Measures the amount of pressure in the lubricating system.
Outside stem and yoke (OS&Y) valve: Valve used to control water supply to a sprinkler system.
Packing: Allows the impeller shaft to pass from outside of the pump to the inside, while maintaining an airtight seal.
Parallel: Capacity position in which each impeller works independently into the discharge.
Pendant: Type of sprinkler head which is suspended from the water pipe.
Percentage: Dividing up into 100 equal parts.

Piston pump: Positive-displacement pump using a piston to develop the movement.
Pitot gage: Measures velocity head at the discharge of a nozzle.
Poppet valve: Valve on the American pump which controls the flow of water into the pump.
Positive-displacement pump: The volume of space within the pump will determine the amount of water which the pump can deliver on one stroke or revolution.
Positive pressure: Pressure above atmospheric.
Post Indicator (PI): Type of water control valve for a sprinkler system.
Potential energy: Energy which has been stored.
Power: Number of times a number is multiplied by itself.
Preaction system: Sprinkler system which provides a warning alarm before the sprinkler head temperature is reached.
Pressure: Force per unit area.
Pressure reducer: Installed on a standpipe to prevent excessive pressures from being supplied to a hand line.
Primary main: Large main which brings water from the source or water treatment plant to the area to be served.
Priming: Process of evacuating the air from the pump so that atmospheric pressure can force water in.
Product: Result of multiplying the multiplicand by the multiplier.
Proper fraction: Fraction whose numerator is smaller than its denominator.
Proportion: Equality of two ratios.
Quotient: Resultant answer.
Radiator fill valve: Permits pump water to directly enter the radiator.
Rate-of-rise system: System which will perform other fire protection tasks (close fire doors, open ventilators, etc.) as well as activate the sprinkler system.
Relay: Movement of water from a pumper at a water source to additional pumpers until the water reaches the fireground.
Ratio: Comparison of like quantities.
Relief valve: Prevents excess pressure on discharge lines by bypassing water from discharge to intake.
Remainder: Extra amount if the quotient is not a whole number.
Residual pressure: Pressure remaining once water has begun flowing.
Retard chamber: Prevents activation of a sprinkler system due to minor water pressure fluctuation.
Riser: Carries water from the intake throughout the building for either standpipe or sprinkler operation.
Root: Determines what number, when multiplied by itself will give the original numbers.
Rotary gear pump: Rotary pump using two gears to move the water.
Rotary pump: Positive-displacement pump using rotary devices to develop water movement.
Rotary vane pump: Rotary pump using metal rods to slide out and seal against the pump housing.
Secondary main: Intermediate size pipes that supply large sections of the service area.
Series: Pressure position in which the first impeller's discharge is fed to the eye of the second impeller which then discharges the water from the pump.
Service test: Pumper test performed in station to determine if the pumper can deliver its rated volumes and pressures.

Shear force: Sideways force.
Shrouds: Sides of the impeller which confine the water.
Siamese connection: Exterior connection for the standpipe system.
Single-acting pump: Type of piston pump that only discharges water while the piston moves in one direction.
Slippage: Backward movement of water from the discharge to the intake.
Speedometer: Indicates speed of apparatus in miles per hour.
Stages: Number of impellers mounted on a common shaft.
Static pressure: Pressure when water is not moving.
Steamer connection: Large discharge on a hydrant.
Subtrahend: Number to be subtracted.
Supervised system: Sprinkler or standpipe system that automatically transmits an alarm when water flows to a place which maintains the status at all times.
Tachometer: Indicates speed of engine crankshaft in revolutions per minute.
Tandem pumping: Connecting two pumpers for a short relay.
Tensile force: Pull apart force.
Term: Combination of symbols between plus or minus signs.
Torque: Ability of the engine to produce rotation at a given speed.
Transfer valve: Selector for series or parallel operation on a multistage pump.
Transposing: Moving a factor from one side of the equation to the other and changing the sign.
Turbulent flow: Water moving with a swirling action.
Upright: Type of sprinkler head mounted upright on the water pipe.
Vacuum primer: Uses the vacuum created by the intake manifold to remove air from the pump.
Vanes: Guides inside the impeller which direct the water to the edge.
Vapor pressure: Pressure created as a confined liquid changes to a gas.
Volute: Gradually increasing discharge waterway.
Water flow alarm: Audible flow alarm on the exterior of a building to indicate a water flow from either the sprinkler or standpipe system.
Water hammer: Shock loading on hose, couplings, and pump due to the sudden stopping of water.
Water horsepower: Amount of work which can be performed by a pump.
Water temperature gage: Indicates the temperature of the waer in the engine cooling system.
Wear rings: Prevents discharge water from returning to the eye of the impeller.
Wet barrel hydrant: Barrel of hydrant always contains water.
Wet pipe sprinkler system: Sprinkler pipes always contain water.

Index

A
Absolute pressure, 51
Accelerators, in sprinkler systems, 218, 220
Acceptance tests, for pumps, 234
Addition
  in algebra, 25
  of decimal fractions, 18
  of fractions, 11
  of numbers, 2
Ahrens-Fox piston pumper, 81
Air brakes, 130
Air chamber
  for piston pumps, 83
  for rotary pumps, 89
Air horns, 130
Air pressure gage, 130
Alley dock driving test, 250
Algebra
  addition in, 25
  division in, 27
  and equations, 29
  grouping symbols in, 28
  multiplication in, 26
  subtraction in, 26
  symbols in, 24
Allison HT-70 transmission, 126-127
Alternator, and ammeter readings, 130
Altitude
  and atmospheric pressure, 51
  and drafting water, 184-185
American dual-impeller centrifugal pump, 100-101, 178
American Insurance Association, 196
American LaFrance centrifugal pump, 96-100
American LaFrance governor, 158-160
American LaFrance primer, 170-171
American LaFrance relief valve, 147-148
American model governor, 161-162
Ammeter, 130
Arabic numbers,
Associative rule
  of addition, 3
  of multiplication, 5
Atmospheric pressure, 50
Atoms, 252
Automatic governor, 167-168
Auxiliary cooling system, 134-136
Average velocity, formula for, 261-263

B
Back pressure, 51
  rule of thumb for, 63
Baffles of tanker, 226
Balancing cylinder, in governor, 158
Base (mathematical), 38
Battery, 130
Bearings, of centrifugal pump, 94
Bernoulli's Law, 259-260, 263
Bleed line, of American LaFrance relief valve, 148
Booster tank, 137-138
Bourdon tube, 132,
Braces, in algebraic operation, 28
Brackets, in algebraic operation, 28
Brake horsepower, 117-119
  calculation of, 118

C
Cab components, 128-131
Cancellation, in multiplying fractions, 14
Cavitation, 158
  and drafting, 188
  at hydrants, 200
  in relay operations, 202
Centrifugal force, principles of, 90-91
Centrifugal pumps
  American dual-impeller model, 100-102
  American LaFrance model, 96-100
  Components of, 91-96
  Hale two-stage model, 103-107
  history of, 78
  operating procedures for, 186-187
  pressure gage, 131-134
  priming of, 177-181
  Seagrave model, 107-108, 111-113
  Waterous two-stage model, 109, 113-116
Certification tests, 235-237
Chain drive, 125-126
Chassis, of tanker, 226
Check valve, in standpipes, 212
Clearance rings, of centrifugal pump, 94
Climatic conditions, and drafting water, 184-185
Clover leaf pumps, 87
Clutch arrangements, 123-124
Coefficient of discharge, 64, 65, 66
Cold weather, and governor maintenance, 164, 166
Common fractions
  decimal conversion of, 17
  defined, 9
  reducing of, 10
Commutative rule
  of addition, 3
  of multiplication, 5
Compound fraction, 9
Compound gage, 131
Compressive forces, 252-253
Conservation of matter, and water flow, 258-259
Control valve, for sprinkler systems, 218
Crankshaft, power transmission from, 123-124
Ctesibius, 75
Cube root, 39

D
da Vinci, Leonardo, 48
Dead-end mains, 194
Decimal fraction
  addition of, 18
  division of, 21
  multiplication of, 19
  percentages, 36
  rounding off, 22
  subtraction of, 19

277

Deck guns, 60, 70
Deluge gun, for service testing, 240
Deluge sprinkler systems, 217
Denominator, defined, 9
Diesel engines, optimum rpms for, 96
Diminishing clearance driving test, 250-251
Direct proportion, defined, 34
Discharge, of tanker, 226
Discharge calculations
    formula, 64
    from fog nozzles, 68, 69
    from smooth bore nozzles, 64, 65
Discharge gages, 134
Discharge pressure line, of American LaFrance relief valve, 148
Discharge valves, 136
Distance, in relay operations, 202
Distributive rule, of multiplication, 5
Dividend, 6
Division
    in algebra, 27
    of fractions, 15
    of numbers, 6
Divisor, 6
Double-acting piston pump, 85
Drafting water (see water supply)
Drain, for discharge/intake lines, 137, 148
Drive shaft, power transmission from, 125, 127
Driving, safe, 247-248
Dry barrel hydrant, 190-191
Dry-pipe valve, in sprinkler system, 217, 218-219
Dry prime test, 243-244
Dynamometer, 119
Dyne, defined, 44

E

Efficiency
    calculation of, 120-122
    single-stage vs. two-stage, 121-122
Electrical system, 130
Elevated storage tank, 192
    for standpipe systems, 213
Elevating platforms, 72
Energy, for hyraulic system, 259-260
Engine gages, 128-130
Engine hours meter, 129
Engine pressure
    calculation of, 67
    friction loss, 67
    importance of,
    relay operations, 205-208
    sprinkler system, 223-224
    standpipe operations, 214-217
Equations, basic procedures for, 29
Exhaust primer, 179-181
Exhauster, in sprinkler system, 218-220
Exponents, 38
Extremes, in proportion, 34
Eye of impeller, 91-94

F

Firefighting, history of, 47, 75
Flinger, ring, 94

Flow
    and conservation of matter, 258-259
    formula for, 64
    and venturi meter, 263
Flow pressure, 52
Fluids
    defined 47-48 252-253
    forces and, 252-153
    properties of, 253-255
Foam sprinkler systems, 217-218
Fog nozzle, 68, 69
    and pressure control, 142-143
    and water flow quantity, 56
Force, categories of, 252
Formula, defined, 25
Fractions (See also decimals)
    addition of, 11
    divison of, 15
    multiplication of, 14
    reducing of, 10
    subtraction of, 13
    types of, 9
Frangible bulb sprinkler head, 222
Freeman, John R., 48
Freeman formula, 266
Friction loss
    and flow type, 54
    and hose, 54, 55
    hand method for, 61
    in devices, 60
    in parallel lines, 60
    in unequal lengths, 60
    other than 2½-inch hose, 58
    in standpipe operations, 215-217
    for 2½-inch lines, 56
    and water flow,
    small flows, 57
Frontinaus, 47
Fuel gage, 129
Fusible link sprinkler heads, 222

G

Gases, properties of, 253
Gasoline engines, history of, 78
Gage pressure, 50
Gages (see specific gage; engine gages; pump gages)
Generator, and ammeter readings, 130-131
Gooseneck engine, 76
Governors
    American LaFrance model, 158-161
    American model, 161-162
    automatic, 167-168
    Hale model, 162-163
    principles of, 158
    Seagrave model, 164-166
    Waterous model, 166-167
Gram, defined, 44
Gravity, and fluid velocity, 260
Grid, 192-193
Gross brake horsepower, 117-119
Ground waters, as supply source, 192
Grouping symbols, in algebra, 28

H
Hale governor, 162-163
Hale priming pump, 172-174
Hale relief valve, 151-153
Hale single-stage pump, 107, 109-110
Hale two-stage pump, 103, 105, 106-107, 108-109
Hard sleeve
  and hydrant connection, 197
  testing of, 243
Head, defined, 51
Height and pressure, 67
High-pressure hydrant, 194
Hosepower of pump drives, 117-119
Hose
  and friction loss,
  history of, 76
  for relay operations, 203
  for standpipes, 212-213
  for tankers, 226-227
Hose tenders, history of, 76
Humidity, and drafting water, 185
Hydrant pressure, 52
Hydrants
  and pumper operation, 196-198
  and water distribution system, 192-194
  characteristics of, 189-191
  estimating available flow, 198-200
  inspection/maintenance, 195-196
  testing, 196
Hydraulics
  calculations, 56
  history of, 47
  for relay operations, 205-208
Hydrokinetics, defined, 47
Hyrostatics, defined, 47
  principles of, 48

I
Impeller, in centrifugal pump, 91, 93, 94
Improper fraction
  defined, 9
  reducing of, 11
Indirect proportion, 35
Intake, of tanker, 226
Intake gage, 134
Intake valves, 136
International Association of Fire Engineers, 234
Inverse proportion, defined, 34

J
John Bean relief valve, 148-149

K
Kinetic energy, defined, 259

L
Ladder pipes, 142
Laminar flow, and friction loss, 53
Lift pump, 80
Lifting, of water, 182-184
Line gage, 134
Liquids, properties of, 253
Liter, defined, 44
Looped water main, 194
Lowest common denominator, 11

M
Mathematics
  Basic operations in, 1
  and fractions, 9
Means, in proportion, 34
Mechanical energy, defined, 259
Meter, defined, 44
Metric system, 43
Minuend, 5
Mixed number
  defined, 10
  dividing of, 16
  subtracting of, 13
Molecules, defined, 252
Multiple-bucket carrier, 76
Multiplicand, 3
Multiplication
  in algebra, 26
  of fractions, 14
  of numbers, 3
Multiplier, 3

N
National Board of Fire Underwriters, 234
National Fire Protection Association, 234
  Standard 1901 text of, 234-235
Needle valve, 134
Negative numbers, subtraction of, 26
Negative pressure, 52
Net brake horsepower, 117-119
Net engine pressure, 52
Net pump pressure, calculation of, 237-239
Newsham engine, 75
Newton (dyne), defined, 44
Nonpositive displacement pump (see centrifugal pump)
Normal operating pressure, 52
Nozzle
  and pressure control, 141-143
  in service tests, 240
  and water flow quantity, 64, 65, 68, 69
Nozzle pressure, 56
Numerator
  of decimal fraction, 17
  defined, 9

O
Odometer, 129
Oil pressure gage, 129
"On the motion and measurement of water," 47-48
One-dimensional flow, 256
  measurement of, 256-257
Opposite alley pull-in test, 250
Overheating, of pump, 138

P
Packing, of centrifugal pump, 94
Parallel lines,
  friction loss in
Parentheses, in algebraic operation, 28
Percentages, principles of, 36
Philadelphia-style hand pumper, 77
Pi, 25
Piano pumper, 77
Pilot valve, with relief valve, 145-147

Pipes, for distribution and storage, 192-194
Piston pumps, 80
   air chamber for, 83
   discharge calculations for,
   double-acting type, 85
   lift type, 80
   multiple cylinder type, 85
   pressure type, 82
   slippage, 86
Pitot, Henri, 48
Pitot, gage, 263-264
   for hydrant testing, 196
   for service test, 240, 242
Portable pumps, 230-232
   and priming, 169
Portable tanks, 229
Positive displacement pumps (see also piston pumps; rotary pumps),
Positive pressure, 52
Potential energy, defined, 259
Power, transmitting of, 123-125
Power takeoff, 127
Powers (mathematics), principles of, 38
Preaction sprinkler systems, 217
Pressure
   and height, 63
   and water, 64
   definition of, 49
   types of, 50
   vs. force, 49
Pressure control, 144
   and governors, 158-167
   and nozzles, 141-143
   and relief valves, 144-157
Pressure gage
   for pumps, 130
   in service tests, 240, 242
Pressure piston pump, 82
Pressure reducer, in standpipes, 212
Pressure tanks, for standpipes, 213
Priming devices (see also rotary priming pumps)
   exhaust type, 179-181
   testing of, 243-244
   vacuum type, 177-179
Prony brake, 129
Proper fraction
   defined, 9
   dividing of, 15
   subtraction of, 12
Proportion, principles of, 34
Pump discharge pressure, 52
Pump drives
   and horsepower, 117-119
   types of, 123-127
Pump gages, 131-134
Pump speed, 122-123
Pumper, driving the, 248-249
Pumps and pumpers (see also centrifugal pumps; piston pumps; pump drives; rotary pumps)
   acceptance testing of, 234
   for tankers, 226
   gasoline, 78
   hand, 75, 77
   history of, 75
   portable, 230-233
   positive displacement type, 80
   relay capabilities of, 202-203
   service testing of, 239-240
   steam, 78
   transmitting power to, 123-127

Q
Quotient, 7

R
Radiator fill valve, 135-136
Radical sign, 39
Rapid water, 56
Rate-of-rise sprinkler system, 217
Rating, of tankers, 227
Ratio
   principles of, 33
   and proportion, 34
Reducing, of fractions, 10
Relay operations
   advantages of, 201-202
   factors in, 201-204
   hydraulics of, 205-208
   pumper operation in, 204-205
Relief valves
   American LaFrance model, 147-148
   basic principles of, 144-145
   Darley model, 149-150
   Hale model, 151-153
   John Bean model, 148-149
   operation of, 157
   simple type, 145
   Thibault model, 153, 154-155
   Waterous model, 154, 155-157
   with pilot valve, 145-147
Remainder (division), 7
Residual pressure, 52
Restricted pressure line, of American LaFrance relief valve, 148
Retard chamber, in sprinkler system, 221
Risers
   in sprinkler system, 221
   in standpipes, 212
Rome, 47
Roof outlet, for standpipes, 213
Roots (mathematical), principles of, 39
Rotary gear pumps, 87
Rotary priming pumps, 169-170
   American LaFrance model, 170-171
   Darley model, 171-172
   Hale model, 172-174
   Seagrave model, 174-175, 176
   Waterous model, 175-177
Rotary pumps, 87
Rotary vane pump, 89
Rubber hose lining, and friction loss, 55

S
Seagrave governor, 164-166
Seagrave pump, 107-108, 111-113
Seagrave rotary vane priming pump, 174-175, 176
Separate engine method, of power transmission, 123
Serpentine driving test, 249-250
Service test, for pumps, 239-240

Shear force, defined, 252, 253
Shrouds, of impeller, 91
Siamese connection
   for sprinkler systems, 218
   for standpipes, 211
Simple (common) fraction, defined, 9
Simple relief valve, 145
Single-stage pump, efficiency of, 121-
Sliding collar drive, 125
Slippage
   in piston pumps, 86
   in rotary pumps, 88
Smooth nozzles, and water flow quantity, 64
Soft sleeve, and hydrant connection, 197
Solids, properties of, 252
Specific gravity, and battery, 130-131
Speedometer, 129
Sprinkler heads, 221-223
Sprinkler systems
   components of, 218-222
   engine pressure for, 223-224
   operating procedures for, 223
   types of, 217-218
Square root, 39
Squirrel-tailed pumper, 77
Stages, of centrifugal pump, 94-95
Standpipes
   components of, 211-213
   engine pressure for, 214-215
   operating procedures for, 213-214
   types of, 210
Static pressure, 52
Steam pumpers, history of, 78
Stevenius, 48
Straight-line driving test, 251
Straight-tipped ground nozzles, and pressure control, 141-142
Straight-tipped ladder pipe, and pressure control, 142
Strainer, in governor, 162
Subscript, in algebra, 25
Subtraction
   in algebra, 26
   of fractions, 12
   of numbers, 5
Subtrahend, 5
Super pumper, 79
Surface waters, as supply source, 192
System International (SI), 43

T
Tachometer, 129, 241
Tank fill valve, 137-138
Tanker
   construction factors of, 226-227
   operating from, 228-229
   rating of, 227
Temperature
   and drafting water, 184-185
   and sprinkler operation, 222-223
Tensile force, defined, 252
Terrain, in relay operations, 204
Testing
   acceptance, 234
   of pumps, 234-237
   service, 239-240

Thibault model relief valve, 153-154
Three-dimensional flow, 256
Throttle linkage clutch, in governor, 159
Time, and fluid velocity, 261
Torque, 119
Training, driver, 249-251
Transfer valve, of centrifugal pump, 95-96
Transposing, in equations, 30
Turbulent flow, 53
Turn-around driving test, 251
$2\frac{1}{2}$-inch lines, friction loss in, 56
Two-stage pump, efficiency of, 121-122

U
Underwriters' formula, 269
Underwriters' Laboratories, 234-235
Unit symbols, for metric system, 43

V
Vacuum primer, 177-178
Vacuum-type transfer, in Hale pumps, 107
Vanes, of impeller, 91
Vapor pressure, and drafting water, 185
Velocity head, 259
Velocity of fluid
   formula for, 261-262
   and pitot gage, 263-264
Vents, of tanker, 226
Venturi, G.B., 48
Venturi meter, 263-264
Vitruvius, 47
Voltage regulator, and ammeter readings, 130
Volute, from impeller, 94

W
Water (see also water supply)
   one-dimensional flow of, 256
   pressure and, 50-52, 254-255
   properties of, 253-254
Water discharge, formula for 64
   Water distribution system and hydrants, 192-194
Water-flow alarm, 219-220
Water hammer, 143
Water horsepower, 119-121
Water supply
   drafting of, 182-188
      and cavitation, 188
      and climatic conditions, 184-185
      and lifting, 182-184
      common problems in, 187-188
      operating procedures for, 186-187
   for sprinkler systems, 223
   for standpipe systems, 213
   history of,
Water tank, 137-138
Water temperature gage, 130
Waterous primer, 175-177
Waterous governor, 166-167
Waterous relief valve, 155-157
Waterous two-stage pump, 109, 113, 114-116
Wear rings
   of centrifugal pump, 94
Wet barrel hydrant, 190

Contributors

The author gratefully acknowledges the materials and help given by the following:

Dave Thomas of the Waterous Company for use of pump efficiency curves. (Reprinted by special permission of the Technical Publishing Corporation from the March 1962 issue of Fire Engineering.)
Akron Brass Company
American Fire Apparatus Company
American LaFrance
American Museum of Fire Fighting, Hudson, N.Y.
Automatic Sprinkler Corporation of America
City of Cincinnati Department of Safety
Cincinnati Historical Society
W. S. Darley Company
Elkhart Brass Mfg. Co., Inc.
Fire Fighting Museum of the Home Insurance Co.
Fire Research Corporation
FMC Corporation, John Bean Division
Hale Fire Pump Company
Ideas Unlimited
MC Products Inc.
Morgan Equipment
Mueller Company
National Fire Protection Association
Reliable Automatic Sprinkler Co., Inc.
Seagrave Fire Apparatus, Inc.
Span Instruments
Pierre Thibault (1972) Ltd.
Uniroyal Plastic Products
Waterous Company

Fire Engineering Staff
Thomas F. Brennan, Editor
Jacqueline Cox, Associate Editor

Acknowledgements

During my fire service career, there have been many individuals who have assisted me in learning about pumps and hydraulics. While it would be impossible to acknowledge them all, there are a few who need to be recognized.

My great interest in pumps was kindled by John Hoglund of the Maryland Fire and Rescue Institute who gave me the responsibility for conducting the Maryland Pump Schools. As part of that responsibility I had the opportunity to learn from Mike Waldoch and C. R. "Skip" Shaffer of Waterous Company and Robert Barraclough then of Hale Pumps and now with Emergency One.

I have also had the opportunity over the years to conduct hydraulics training classes in many different places. Each of these classes enabled me to add information to the textbook. As a result of these classes I would like to thank the following individuals who aided in the writing of this text:

Bill Hagevig, Alaska Fire Training
Lieutenant Warren Cummings, Fairbanks, Alaska, Fire Department
Chief Tom Opie, Barrow, Alaska Fire Department
Gene Carlson, IFSTA

Finally, the hardest part of preparing a textbook is to change the handwritten material into a readable document. This task was ably handled by Darlene Skelton. Her editing and accurate typing were extremely helpful in changing manuscript into final text and is gratefully acknowledged.